SpringerBriefs in Space Life Sciences

Series Editors

Günter Ruyters
Markus Braun
Space Administration
German Aerospace Center (DLR)
Bonn, Germany

The extraordinary conditions of space, especially microgravity, are utilized for research in various disciplines of space life sciences. This research that should unravel – above all – the role of gravity for the origin, evolution, and future of life as well as for the development and orientation of organisms up to humans, has only become possible with the advent of (human) spaceflight some 50 years ago. Today, the focus in space life sciences is 1) on the acquisition of knowledge that leads to answers to fundamental scientific questions in gravitational and astrobiology, human physiology and operational medicine as well as 2) on generating applications based upon the results of space experiments and new developments e.g. in non-invasive medical diagnostics for the benefit of humans on Earth. The idea behind this series is to reach not only space experts, but also and above all scientists from various biological, biotechnological and medical fields, who can make use of the results found in space for their own research. SpringerBriefs in Space Life Sciences addresses professors, students and undergraduates in biology, biotechnology and human physiology, medical doctors, and laymen interested in space research. The Series is initiated and supervised by Dr. Günter Ruyters and Dr. Markus Braun from the German Aerospace Center (DLR). Since the German Space Life Sciences Program celebrated its 40th anniversary in 2012, it seemed an appropriate time to start summarizing – with the help of scientific experts from the various areas - the achievements of the program from the point of view of the German Aerospace Center (DLR) especially in its role as German Space Administration that defines and implements the space activities on behalf of the German government.

More information about this series at http://www.springer.com/series/11849

Christine E. Hellweg • Thomas Berger •
Daniel Matthiä • Christa Baumstark-Khan

Radiation in Space: Relevance and Risk for Human Missions

 Springer

Christine E. Hellweg
Department of Radiation Biology, German
Aerospace Center (DLR), Institute of
Aerospace Medicine
Köln, Nordrhein-Westfalen, Germany

Daniel Matthiä
Department of Radiation Biology, German
Aerospace Center (DLR), Institute of
Aerospace Medicine
Köln, Nordrhein-Westfalen, Germany

Thomas Berger
Department of Radiation Biology, German
Aerospace Center (DLR), Institute of
Aerospace Medicine
Köln, Nordrhein-Westfalen, Germany

Christa Baumstark-Khan
Department of Radiation Biology, German
Aerospace Center (DLR), Institute of
Aerospace Medicine
Köln, Nordrhein-Westfalen, Germany

ISSN 2196-5560   ISSN 2196-5579   (electronic)
SpringerBriefs in Space Life Sciences
ISBN 978-3-030-46743-2   ISBN 978-3-030-46744-9   (eBook)
https://doi.org/10.1007/978-3-030-46744-9

This Springer imprint is published by the registered company Springer Nature Switzerland AG.
The registered company address is: Gewerbestrasse 11, 6330 Cham, Switzerland

Foreword

Up to now, the ten books published in our series "Springer Briefs in Space Life Sciences" all have mainly focused on the effects of altered gravity conditions, especially microgravity, on living systems—from cells and microorganisms, plants and animals up to humans. However, also other environmental conditions are changed in space, above all the radiation field.

In fact, solar and galactic cosmic radiation are considered the main health hazard for human exploration and colonization of the solar system. Radiation risk is characterized by a high uncertainty and lack of simple countermeasures. Most of the uncertainty on space radiation risk is associated with the poor knowledge of biological effects of solar and cosmic rays. This creates the need for investigations into biological effects of space radiation, in order to allow more accurate risk assessments, which in turn could lead to more accurate planning of countermeasures. Moreover, results from numerous radiation measurements indicate that not only the radiation levels are increased in space, but also the nature of the radiation field changes, especially with regard to the presence of high energy heavy ions.

The different radiation fields exert various negative consequences on humans, such as DNA damage, carcinogenesis, central nervous system effects, degeneration of tissues, and other health effects. In addition, these risks from radiation exposure may be influenced by other spaceflight factors like microgravity and environmental contaminants; interaction of radiation and microgravity especially on the cellular level has been frequently demonstrated. From this it is clear that a mission—for instance—to Mars will not be feasible unless improved shielding or other effective (biological) countermeasures have been developed, and it is also obvious that our series "Springer Briefs in Space Life Sciences" is not complete without dealing with this important topic.

The authors of this book *Radiation in Space: Relevance and Risk for Human Missions* cover all these important aspects: After a general introduction to the topic, they describe in detail the physics of radiation in space. This includes the description of the different radiation sources present in low Earth orbit and beyond, methods and devices to measure the radiation (the so-called dosimetry) as well as possibilities to model space radiation on Earth and learn about its effects.

Chapter 3 deals with the biological aspect of space radiation, i.e., with the effects of radiation on the various tissues of organisms up to humans. Acute, chronic, and late radiation effects are described, such as those on the central nervous and the cardiovascular systems. These aspects are, of course, also of relevance for people on Earth, radiation therapy being of high importance for patients with cancer treatment.

The same holds true for Chap. 4, in which the authors describe the risk assessment. It becomes obvious that—in spite of radiation measurements in space having been performed for decades—uncertainties still exist leading to discussions on the acceptability of risk. The similarity to the treatment of patients is evident.

Consequently, Chap. 5 covers the development of countermeasures. A wide range of possibilities is described here, ranging from operational planning, shielding, nutritional and pharmaceutical countermeasures up to crew selection.

Finally, the book closes with asking the question: Are we ready for launch? Do we know enough about the risk and about the effectivity of countermeasures? Can we take the responsibility to send humans to Mars knowing about the radiation risk? The answers to these questions certainly are also dependent on the position of the scientists or the reader of this book towards human spaceflight in general and towards exploratory missions in detail. However, history shows that mankind has always pushed its frontiers and moved forward to new horizons. So, in our mind, the question is not, if humans will go to Mars and other distant destinations, but only who and when!

Bonn, Germany   Günter Ruyters
March 2020   Markus Braun

Preface to the Series

The extraordinary conditions in space, especially microgravity, are utilized today not only for research in the physical and materials sciences—they especially provide a unique tool for research in various areas of the life sciences. The major goal of this research is to uncover the role of gravity with regard to the origin, evolution, and future of life, and to the development and orientation of organisms from single cells and protists up to humans. This research only became possible with the advent of manned spaceflight some 50 years ago. With the first experiment having been conducted onboard Apollo 16, the German Space Life Sciences Program celebrated its 40th anniversary in 2012—a fitting occasion for Springer and the DLR (German Aerospace Center) to take stock of the space life sciences achievements made so far.

The DLR is the Federal Republic of Germany's National Aeronautics and Space Research Center. Its extensive research and development activities in aeronautics, space, energy, transport, and security are integrated into national and international cooperative ventures. In addition to its own research, as Germany's space agency the DLR has been charged by the federal government with the task of planning and implementing the German space program. Within the current space program, approved by the German government in November 2010, the overall goal for the life sciences section is to gain scientific knowledge and to reveal new application potentials by means of research under space conditions, especially by utilizing the microgravity environment of the International Space Station (ISS).

With regard to the program's implementation, the DLR Space Administration provides the infrastructure and flight opportunities required, contracts the German space industry for the development of innovative research facilities, and provides the necessary research funding for the scientific teams at universities and other research institutes. While so-called small flight opportunities like the drop tower in Bremen, sounding rockets, and parabolic airplane flights are made available within the national program, research on the ISS is implemented in the framework of Germany's participation in the ESA Microgravity Program or through bilateral cooperations with other space agencies. Free flyers such as BION or FOTON satellites are used in cooperation with Russia. The recently started utilization of Chinese spacecrafts like Shenzhou has further expanded Germany's spectrum of flight opportunities, and discussions about future cooperation on the planned Chinese Space Station are currently underway.

From the very beginning in the 1970s, Germany has been the driving force for human spaceflight as well as for related research in the life and physical sciences in Europe. It was Germany that initiated the development of Spacelab as the European contribution to the American Space Shuttle System, complemented by setting up a sound national program. And today Germany continues to be the major European contributor to the ESA programs for the ISS and its scientific utilization.

For our series, we have approached leading scientists first and foremost in Germany, but also—since science and research are international and cooperative endeavors—in other countries to provide us with their views and their summaries of the accomplishments in the various fields of space life sciences research. By presenting the current SpringerBriefs on muscle and bone physiology we start the series with an area that is currently attracting much attention—due in no small part to health problems such as muscle atrophy and osteoporosis in our modern aging society. Overall, it is interesting to note that the psycho-physiological changes that astronauts experience during their spaceflights closely resemble those of aging people on Earth but progress at a much faster rate. Circulatory and vestibular disorders set in immediately, muscles and bones degenerate within weeks or months, and even the immune system is impaired. Thus, the aging process as well as certain diseases can be studied at an accelerated pace, yielding valuable insights for the benefit of people on Earth as well. Luckily for the astronauts: these problems slowly disappear after their return to Earth, so that their recovery processes can also be investigated, yielding additional valuable information.

Booklets on nutrition and metabolism, on the immune system, on vestibular and neuroscience, on the cardiovascular and respiratory system, and on psycho-physiological human performance will follow. This separation of human physiology and space medicine into the various research areas follows a classical division. It will certainly become evident, however, that space medicine research pursues a highly integrative approach, offering an example that should also be followed in terrestrial research. The series will eventually be rounded out by booklets on gravitational and radiation biology.

We are convinced that this series, starting with its first booklet on muscle and bone physiology in space, will find interested readers and will contribute to the goal of convincing the general public that research in space, especially in the life sciences, has been and will continue to be of concrete benefit to people on Earth.

DLR Space Administration in Bonn-Oberkassel (DLR)

The International Space Station (ISS); photo taken by an astronaut from the space shuttle Discovery, March 7, 2011 (NASA)

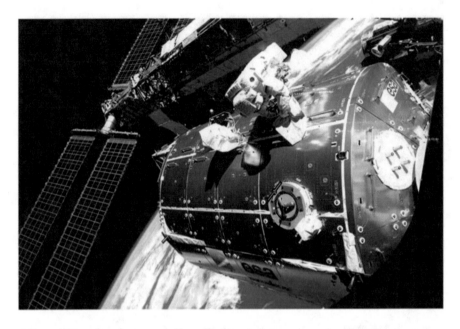

Extravehicular activity (EVA) of the German ESA astronaut Hans Schlegel working on the European Columbus lab of ISS, February 13, 2008 (NASA)

Bonn, Germany   Günter Ruyters
July 2014   Markus Braun

Acknowledgements

Own research described in this book has been made possible by scientific cooperation and by funding by the German Aerospace Center (DLR) grant FuE-Projekt "ISS LIFE" (program RF-FuW, "Teilprogramm 475"; funding of university partners by the DLR Space Agency), by funding of the Helmholtz Association, the European Union, and the National Institute of Radiation Science (NIRS). Doctoral students involved in this research participated in the Helmholtz Space Life Sciences Research School (SpaceLife), German Aerospace Center (DLR; Cologne, Germany), which was funded by the Helmholtz Association (Helmholtz Gemeinschaft) during a period of 6 years (grant VH-KO-300) and received additional funds from the DLR, including the Aerospace Executive Board and the Institute of Aerospace Medicine. The accelerator experiments at the "Grand Accélérateur National d'Ions Lourds" (GANIL), in Caen, France, and the "GSI Helmholtzzentrum für Schwerionenforschung" were supported in part by the European Union (EURONS, ENSAR) and by the European Space Agency (ESA) in the program "Investigations Into the Biological Effects of Radiation" (IBER), respectively. The travel costs to the Heavy Ion Medical Accelerator in Chiba (HIMAC), Japan, were supported in part by the International Open Laboratory of the National Institute of Radiation Science (NIRS) and National Institutes for Quantum and Radiological Science and Technology (QST), Anagawa, Inageku, in Chiba, Japan. We thank Isabelle Testard, Amine Cassimi, Hermann Rothard, Yannick Saintigny, Florent Durantel, and all the physicists of CIMAP (Caen, France) involved in dosimetry and the beam operator team at GANIL for providing us valuable advice and many help given during numerous night shifts at the French Heavy Ion Accelerator GANIL. Michael Scholz, Chiara La Tessa, Ulrich Weber, Insa Schröder, and the beam operators are acknowledged for their valuable help during beam times GSI. The authors would like to thank Teruaki Konishi, Satoshi Kodaira, and Hisashi Kitamura from NIRS and QST, Anagawa, Inageku, in Chiba, Japan, for the support during HIMAC beamtimes. We thank all astronauts and cosmonauts who provided their support for the performance of experiments on board the ISS as well as colleagues from CADMOS, Toulouse, France; DLR-MUSC,

Cologne, and OHB Bremen, Germany as well as all colleagues from the "ESA European Space Agency." Last but not least the authors would like to thank Dr. Gerda Horneck and Dr. Günther Reitz, the former heads of the Radiation Biology Department at the DLR Institute of Aerospace Medicine, for their scientific advice and continuous support of our work.

Contents

About the Authors

Christine E. Hellweg is Head of the Department of Radiation Biology at the Institute of Aerospace Medicine, German Aerospace Center (DLR) in Cologne, Germany. Her Department addresses aerospace-related topics concerning the effects of radiation on humans and the biosphere, as well as characterizing the unique radiation field in space. She studied Veterinary Medicine at the Freie Universität Berlin, where she currently teaches courses on immunology. She has conducted numerous biological experiments at heavy ion accelerators, and is a member of the International Academy of Astronautics (IAA).

Thomas Berger is Head of the Biophysics Working Group, Department of Radiation Biology. He studied physics at the Technical University of Vienna (TUW), Austria and graduated with a PhD in Radiation Physics. His main research interests are in radiation protection, including the development of radiation detectors, and in investigating the radiation load received by humans for missions beyond Low Earth Orbit (LEO). He is currently the Principal Investigator (PI) for the DOSIS 3D experiment on board the International Space Station (ISS) and the Matroshka AstroRad Radiation Experiment (MARE) project, which will accompany NASA's Orion Exploration Mission 1 (EM-1) to the Moon.

Daniel Matthiä holds a PhD in Physics and is a scientist in the Department of Radiation Biology at the Institute of Aerospace Medicine, German Aerospace Center. As an expert in the field of cosmic radiation and dosimetry in spaceflight and aviation, he has previously developed an engineering model for primary galactic cosmic radiation applicable in the field of radiation protection in space and aviation (DLR GCR Model) and the PANDOCA model for the assessment of aircrews' radiation exposure. His main research interest is in the analysis and mitigation of radiation risk from galactic cosmic radiation and solar energetic particles.

Christa Baumstark-Khan was head of the Cellular Biodiagnostics group at the Institute of Aerospace Medicine, Department of Radiation Biology. She holds a professorship (radiobiology) at the Bonn-Rhein-Sieg University of Applied Sciences. She has extensive experience in radiation biology, and studied the role of gravity in DNA repair processes on the second International Microgravity Laboratory (IML-2) mission.

Abbreviations

AD	Alzheimer's Disease
ALFMED	Apollo Light Flash Moving Emulsion Detector
APC	Adenomatous Polyposis Coli
ApoE	Apolipoprotein E
ApoE$^{-/-}$	ApoE-Deficient
ARS	Acute Radiation Syndrome
AT	*Ataxia Telangiectasia*
ATM	*Ataxia Telangiectasia* Mutated
BEIR	Biological Effects of Ionizing Radiation
BER	Base Excision Repair
β	Particle Velocity relative to the speed of light
BFO	Blood-Forming Organs
BfS	"Bundesamt für Strahlenschutz" (Federal Office for Radiation Protection)
CDK	Cyclin-Dependent Kinases
CDKN1A	Cyclin-Dependent Kinase Inhibitor 1A
CHO	Chinese Hamster Ovary
CI	Confidence Interval
CME	Coronal Mass Ejection
CMO	Crew Medical Officer
CNS	Central Nervous System
COX-2	Cyclooxygenase-2
CPDS	Charged Particle Detector
CRAND	Cosmic Ray Albedo Neutron Decay
CtIP	C-Terminal Binding Protein Interacting Protein
DDR	DNA Damage Response
DDREF	Dose and Dose-Rate Reduction Effectiveness Factor
DNA	Deoxyribonucleic Acid
DNA-PK	DNA-Dependent Protein Kinase
DNA-PKcs	DNA-Dependent Protein Kinase, Catalytic Subunit
DSB	Double-Strand Breaks

E	Particle Kinetic Energy Per Nucleon
EM	Exploration Mission, Now: Artemis
EPIC	European Prospective Investigation into Cancer and Nutrition
ERR	Excess Relative Risk
ESA	European Space Agency
ESP	Energetic Storm Particles
EVA	Extravehicular Activity
EV-CPDS	Extravehicular Charged Particle Directional Spectrometer
GADD45α	Growth Arrest and DNA Damage Inducible Gene 45α
GANIL	Grand Accélérateur National d'Ions Lourds
GCR	Galactic Cosmic Radiation or Rays
G-CSF	Granulocyte Colony-Stimulating Factor
GLE	Ground Level Enhancement/Event
GOES	Geostationary Operational Environmental Satellite
GSI	GSI Helmholtzzentrum für Schwerionenforschung GmbH
Gy	Gray
Gy-Eq	Gray-Equivalent
H	Dose Equivalent
HDL	High-Density Lipoprotein
HSC	Hematopoietic Stem Cells
HPRT	Hypoxanthine Phosphoribosyltransferase
HR	Homologous Recombination
IBMP	Institute for Biomedical Problems
ICRP	International Commission on Radiation Protection
ICRU	International Commission on Radiological Units and Measurements
IL	Interleukin
IML-2	International Microgravity Laboratory 2
IRIF	Ionizing Radiation-Induced Foci
ISS	International Space Station
ITS	Interplanetary Transport System
IV-CPDS	Intravehicular Charged Particle Directional Spectrometer
JAXA	Japanese Aerospace and Exploration Agency
JSC	Johnson Space Center
KGF	Keratinocyte Growth Factor
LAR	Lifetime Attributable Risk
$LD_{50/30}$	Dose That Results in 50% Mortality Within 30 Days
LDL	Low-Density Lipoprotein
LDLr	LDL-Receptor
LEO	Low Earth Orbit
LET, L_Δ or L	Linear Energy Transfer
LNT	Linear No-Threshold
LOC	Loss of Crew
LOM	Loss of Mission

LSAH	Longitudinal Survey of Astronaut Health
mBAND	multicolor Banding In Situ Hybridization
MeV/n	Mega Electron Volt Per Nucleon
MORD	Medical Operations Requirements Document
MPCV	Multi-Purpose Crew Vehicle
MRN	MRE11-RAD50-NBS1
MSL	Mars Science Laboratory
mSv	milliSievert
NAS	National Academy of Sciences
NASA	National Aeronautics and Space Administration
NCR	National Cancer Institute
NCRP	National Council on Radiation Protection and Measurements
NER	Nucleotide Excision Repair
NF-κB	Nuclear Factor κB
NGF	Nerve Growth Factor
NHEJ	Non-Homologous End-Joining
NK Cells	Natural Killer Cells
NOAA	National Oceanic Atmospheric Administration
NRC	National Research Council
Nrf2	Nuclear Erythroid-Derived 2-Related Factor 2
NSAIDs	Nonsteroidal Anti-Inflammatory Drugs
NSCR Model	NASA Space Cancer Risk Model
NSRL	NASA Space Radiation Laboratory
$p21^{Cip1,WAF1}$	Protein $21^{\text{CDK2-interacting protein 1, wild-type p53-activated fragment 1}}$ (Encoding Gene: CDKN1A, Cyclin-Dependent Kinase Inhibitor 1A)
PEL	Permissible Exposure Limits
PSD	Positron Sensitive Detector
Q(LET)	Quality Factor
QF	NASA's Space Radiation Quality Factor
RAD	Radiation Assessment Detector
RAM	Radiation Area Monitor
RBE	Relative Biological Effectiveness
R_C	Effective Vertical Cut-Off Rigidity
REID	Risk of Exposure-Induced Death
ROS	Reactive Oxygen Species
RPA	Replication Protein A
RRM2B	Ribonucleotide Reductase p53R2
SAA	South Atlantic Anomaly
SCF	Stem Cell Factor
SEC	Space Environment Center
SEP	Solar Energetic Particle
SOD	Superoxide Dismutase
SOBP	Spread-Out Bragg Peak

SPE	Solar Particle Event
SRAG	Space Radiation Assessment Group
SRHO	Space Radiation Health Officer
SSA	Single Strand Annealing
SSB	Single-Strand Breaks
ssDNA	Single-Stranded DNA
StrlSchG	"Strahlenschutzgesetz" (German Radiation Protection Law)
Sv	Sievert
TBI	Total Body Irradiation
TCR	T-Cell Receptor
TEPC	Tissue Equivalent Proportional Counter
TLD	Thermoluminescent Detector
TLR5	Toll-Like Receptor 5
TNF-α	Tumor Necrosis Factor α
TRL	Technology Readiness Level
UNSCEAR	United Nations Scientific Committee on the Effects of Atomic Radiation
US	United States
V(D)J Recombination	Genetic Recombination of the Variable (V), Diversity (D) and Joining (J) Gene Segments of Immunoglobulin (Ig), and T Cell Receptors (TCR) Genes
XRCC4	X-Ray Repair Complementing Defective Repair in Chinese Hamster Cells 4
XPC	*Xeroderma pigmentosum*, Complementation Group C
Z	Atomic Number

List of Figures

Chapter 1
General Introduction

Abstract Radiation exposure of astronauts during long-term space missions exceeds dose limits for terrestrial occupational radiation exposure by far. In the atmosphere, exposure to cosmic radiation augments with increasing height, resulting in an additional average annual radiation dose to aircrew of 2 mSv in Germany. Leaving the atmosphere and entering space, the radiation exposure enhances further, by the lack of atmospheric and magnetic shielding and by the additional contribution from particles contained in the radiation belts. The complexity of the space radiation field poses a huge challenge for dosimetry and evaluation of its biological effects. Traversals of energetic charged particles of cosmic radiation can even be perceived as light flashes by astronauts after dark adaptation. Countless heavy ion accelerator experiments were and are performed in order to understand the effects of heavy ions at molecular, cellular, tissue, and organismal level and to develop countermeasures.

Keywords Cosmic radiation · Geomagnetic field · Astronauts · Radiation protection regulation · Light flashes · Heavy ion accelerators · Space radiation biology

Traveling through space to new worlds is humankind's dream and visualized in numerous science fiction movies. However, space is not empty as perceived by the eye and high energy charged particles originating from the Sun (solar energetic particles, SEP) or from outside our solar system (galactic cosmic rays, GCR) rush through this vastness, collectively called cosmic radiation (Fig. 1.1). On Earth, the geomagnetic field and the atmosphere provide shielding against this natural radiation source. The GCR interact with the atmosphere, resulting in a shower of primary and secondary particles with a wide range of energies. This leads to an average

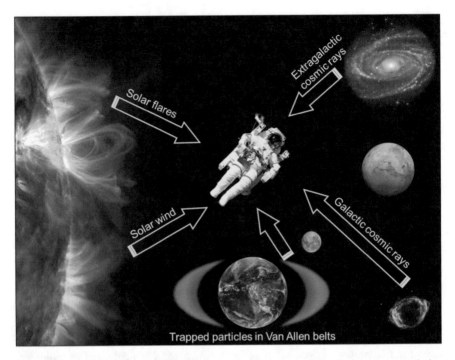

Fig. 1.1 Sources of cosmic radiation that contribute to astronauts' radiation exposure during space missions. The depiction was compiled using the following images: Sun: Coronal Loops in an Active Region of the Sun, Feb. 18, 2014, Image Credit: NASA/Solar Dynamics Observatory; Earth: Satellite View of the Americas on Earth Day, April 23, 2014, Image Credit: NASA/NOAA/ GOES Project; Moon: A full Moon above the South Pacific Ocean, Aug. 13, 2019, iss060e01347, Expedition 60, International Space Station (ISS), Image Credit: NASA; Mars: Valles Marineris: The Grand Canyon of Mars, March 23, 2008, Image Credit: NASA; Galaxy: Spiral Galaxy M81, May 2, 2014, Image credit: X-ray: NASA/CXC/SAO; Optical: Detlef Hartmann; Infrared: NASA/ JPL-Caltech; Supernova remnant: Supernova G1.9 + 0.3, March 30, 2016, Image credit: NASA/ CXC/CfA/S. Chakraborti et al.; Astronaut: Astronaut Bruce McCandless Floating Free (STS-41B), July 28, 2011, Image Credit: NASA

contribution to the annual natural radiation exposure on ground of 0.38 milliSievert (mSv). The dose rate of cosmic radiation rises with increasing height, resulting in higher exposures during a stay on high-altitude mountains or during air travel. At high latitudes around the poles, where the magnetic field lines are perpendicular to the Earth surface, the intensity of the primary charged particles is greater than at lower latitudes. In consequence, aircrew accumulate quite considerable doses of this natural radiation (in average 1.96 mSv per year in Germany), with higher exposures during high latitude flights compared to flight paths over the equator. In Germany, radiation protection for aircrew is regulated by law. Since 2003, the occupational radiation exposure of aircrew has to be assessed, which is mostly realized by using dose calculation software.

Fig. 1.2 Amateur astrophotographer Javier Manteca captured the International Space Station (ISS) as it flew in front of the Moon on 5 February 2020. ©Javier Manteca / ESA; downloaded from https://www.esa.int/ESA_Multimedia/Search?SearchText=astronaut&result_type=images

In low Earth orbit (LEO), at the altitude of the International Space Station (ISS, Fig. 1.2), astronauts are exposed to much higher doses of GCR compared to aircrew and additionally to a larger fraction of primary particles of GCR including heavy nuclei and to the charged particles trapped in the radiation belts (Fig. 1.1).

Interestingly, the occupational radiation exposure of astronauts is regulated by law in Germany only since 2017 with the new radiation protection law ("Strahlenschutzgesetz," StrlSchG) (Bundesrepublik Deutschland 2017) which is based on the European Union Directive 2013/59/EURATOM (Council of the European Union 2014). These regulations follow the recommendations of the International Commission for Radiation Protection (ICRP) to include exposures to natural radiation sources as part of occupational exposure in case of operation of jet aircraft and spaceflight (ICRP 1991, 2007).

The operation of spacecraft in Germany is due for notification in case that astronauts on this spacecraft can be exposed to a cosmic radiation dose of more than 1 mSv per year (StrlSchG § 52). The "Bundesamt für Strahlenschutz (BfS)" (Federal Office for Radiation Protection) is responsible for control of this notification and of compliance with the radiation protection rules (StrlSchG § 185). Upon special notification and justification, the annual occupational exposure limit of 20 mSv (exceptionally 50 mSv for a single year) can be exceeded for astronauts of such German spacecraft (StrlSchG § 52).

These limits are exceeded by a six-month ISS mission. Nevertheless, radiation protection for astronauts has been and is expected to remain a task of the responsible space agencies. This applies certainly for European Space Agency (ESA) astronauts as the respective spacecraft are (currently) not operated by a German organization.

In terms of risk assessment for radiation protection, cosmic radiation poses a huge challenge due to its complexity and the presence of energetic heavy ions. The charged particles can ionize atoms and molecules by pushing electrons out of their shell. Usually, humans have no senses for ionizing radiation. Space radiation is an exception to this rule as some particle traversals through the retina of the eye can be perceived as light flashes. It was predicted by Cornelius Tobias in 1952 that cosmic radiation can interact with the visual system and induce anomalous light perceptions. The Apollo 11 astronaut Edwin E. Aldrin Jr. first reported such light perceptions and spontaneous observations were documented for Apollo 11 and 12 missions (Akatov et al. 1996). During Apollo 15–17, the appearance of the light flashes in the form of stars, streaks, clouds, etc. was noted during observation sessions after dark adaptation. In order to correlate the light flashes to charged particle traversals, dedicated detectors were constructed and used during space missions, starting with the Apollo Light Flash Moving Emulsion Detector (ALFMED) that was applied during Apollo 16 and 17 (Akatov et al. 1996). MIR and ISS astronauts also counted light flashes and reported an increase in number during South Atlantic Anomaly (SAA, Chap. 2) passage (Narici 2008).

Space radiation dosimetry cannot be achieved with devices designed for radiation qualities on Earth (e.g., X-rays, α-, β-, γ-radiation, neutrons), but requires specialized dosimeters operating in passive or active ways. Furthermore, while traveling through the human body, energy deposition and the radiation quality of the charged particles change along the track of the particle (ICRP 2007). Therefore, the depth dose distribution in the human body is a major pillar for space radiation risk assessment. The radiation field in space and approaches for dosimetry and depth dose determination are discussed in Chap. 2.

Simulation of the whole space radiation field on Earth using heavy ion accelerators is impossible. Parts of the space radiation field (e.g., a combination of protons, helium ions, and a heavy ion) can be simulated with high technical effort at high energy heavy ion accelerators. Therefore, experiments performed to understand the biological effects of energetic charged particles are predominantly performed with single ions. Hundreds of researchers contributed to current understanding of radiobiological effects of space radiation, performing thousands of night shifts at heavy ion accelerators with diverse biological systems including bacteria, cell and organ cultures, rodents and larger animals, or even self-experiments (Fig. 1.3). Chapter 3 can give only a glimpse of past and present space radiobiology with a focus on molecular mechanisms, relative biological effectiveness of space-relevant radiation qualities and effects at the organismal level.

Determination of exposure during space missions and quantification of disease risks are the prerequisites for radiation risk assessment which is described in Chap. 4. A Mars mission would exceed the acceptable risk level, therefore, countermeasures have to be developed and tested in order to keep the risk for astronauts' health

Fig. 1.3 Radiobiological experiments at heavy ion accelerators. Left: 25 cm² cell culture flasks and slide flasks filled with medium on the conveyor belt at GSI Helmholtzzentrum für Schwerionenforschung GmbH in Darmstadt, Germany. Right: Irradiation of mammalian cells in 12.5 cm² for determination of cell survival and in slide flasks for immunofluorescence staining at the Grand Accélérateur National d' Ions Lourds (GANIL) in Caen, France. In the lower picture, a holder for degraders to modulate beam energy was installed in front of the beam exit window. © Christine E. Hellweg, DLR

in an acceptable range. The current status of countermeasure research is summarized in Chap. 5.

The health risks provoked by space radiation exposure remain an unsolved question for long-term human space exploration missions. Approaches to answer the open question are described in Chap. 6.

References

Akatov Y, Arkhangelsky V, Petrov V, Trukhanov K, Avdeev S, Ozerov Y, Popov A, Fuglesang C (1996) Biomedical aspects of light flashes observed by astronauts during spaceflight. ESA Special Publication, Paris

Council of the European Union (2014) Council Directive 2013/59/Euratom of 5 December 2013 laying down basic safety standards for protection against the dangers arising from exposure to ionising radiation, and repealing Directives 89/618/Euratom, 90/641/Euratom, 96/29/Euratom, 97/43/Euratom and 2003/122/Euratom. Off J Eur Commun 13:1–73

Bundesrepublik Deutschland (2017) Gesetz zum Schutz vor der schädlichen Wirkung ionisierender Strahlung (Strahlenschutzgesetz–StrlSchG). Strahlenschutzgesetz vom 27. Juni 2017 (BGBl. I S. 1966), das zuletzt durch Artikel 3b des Gesetzes vom 28. April 2020 (BGBl. I S. 960) geändert worden ist. www.gesetze-im-internet.de/strlschg.StrlSchG.pdf

ICRP (1991) ICRP Publication 60. 1990 Recommendations of the International Commission on Radiological Protection. Ann ICRP 21(1–3):1–201, Smith H (Ed.), Pergamon Press, Oxford, New York, Frankfurt, Seoul, Sydney, Tokio

ICRP (2007) ICRP Publication 103. The 2007 Recommendations of the International Commission on Radiological Protection. Ann ICRP 37(2–4):1–332, Valentin J (Ed.), Elsevier, Orlando, Amsterdam, Tokyo, Singapore

Narici L (2008) Heavy ions light flashes and brain functions: recent observations at accelerators and in spaceflight. New J Phys 10(7):075010

Chapter 2
Radiation in Space: The Physics

Abstract The radiation field in space is highly variable in time and space. Different sources contribute to the total exposure. In interplanetary space, the field is dominated by the omni-present galactic cosmic radiation (GCR) and sporadic solar particle events (SPE) can contribute. On the International Space Station (ISS) in low Earth orbit (LEO), on the other hand, the contribution of SPE can be neglected and GCR are modulated along the station's trajectory due to the shielding effect of the geomagnetic field against charged particles. On planetary surfaces, for instance, on Mars, albedo particles from underground and secondary particles from interactions with the atmosphere, if present, are added to the radiation field. Secondary particles, especially neutrons, can contribute significantly to the exposure. In all cases, the field can be further modified by the potential shielding environment and the resulting particle fluxes lead to the exposure of humans under the given conditions. The exposure is calculated as the energy deposition in tissue weighted with corresponding quality factors or relative biological effectiveness and organ weighting factors. In most cases, if measured, the dose rate is determined from the energy deposition in silicon detectors and corresponding corrections have to be applied to estimate the dose in tissue. Additionally, self-shielding of the body has to be taken into account if organ doses are determined.

Keywords Galactic cosmic rays (GCR) · Solar particle events (SPE) · Van Allen Belts · Absorbed dose · Dose equivalent · Effective dose equivalent · Anthropomorphic phantoms · Passive detectors · Active detectors · Radiation exposure

2.1   The Radiation Field

Every human being is permanently exposed to natural radiation. For the majority of humankind, the largest part of this radiation stems from naturally occurring ambient radioactive nuclei in the soil or their decay products. The contribution of cosmic radiation to the total natural exposure is approximately 15% on average (UNSCEAR (2000) gives an average value for the yearly effective dose from cosmic radiation of 380 µSv and a total of 2.4 mSv). The actual value depends among others on the altitude, the local soil composition, and diet.

Under certain conditions, however, the contributions of radiation which originates not from Earth but from space increases and may become the only relevant factor. The sources of this radiation are threefold: atomic nuclei that are accelerated to extremely high energies at extra-heliospheric sources in the galaxy, the galactic cosmic radiation (GCR), protons and electrons that are accelerated in solar flares or coronal mass ejections close to the visible solar surface or in interplanetary space, the solar energetic particles (SEP); and protons and electrons temporarily trapped in the magnetic field of Earth in the radiation belts. The nature of this radiation differs from radiation sources found on Earth in energy and composition. Heavy ions in the GCR have a high biological effectiveness, and they reach energies at which it is impossible to shield against them under the current technical constraints in human spaceflight. The interaction of the primary radiation with shielding material creates a complex field of secondary radiation containing an increasingly large fraction of neutrons. Energies of particles from the radiation sources encountered in space usually have much greater energies than typical terrestrial sources. Alpha particles (fully ionized He nuclei) originating from radioactive decay, for instance, have kinetic energies on the order of MeV, the energies of those from GCR exceed tens and even hundreds of GeV which allows them to penetrate shielding that is orders of magnitude thicker compared to what is needed to shield alpha particles from radioactive decay.

Figure 2.1 gives an overview over the energy regimes of different sources of radiation in space. Sources that are potentially relevant for the exposure to ionizing radiation are trapped protons and electrons (during extravehicular activities) in low Earth orbit (LEO), protons of solar origin, and galactic cosmic radiation.

2.1.1   Galactic Cosmic Rays

Galactic cosmic rays (or synonymously used: galactic cosmic radiation (GCR)) consist of highly energetic particle radiation that enters the heliosphere from interstellar space and originates at galactic sources like supernova remnants (Blasi 2013). The intensity of GCR in the interstellar space is considered to be effectively constant over time and outside the heliosphere, the intensity of GCR is described by the local interstellar spectra. On their way to a given location in the heliosphere, the shape of the GCR spectra changes through the interaction of the charged particles

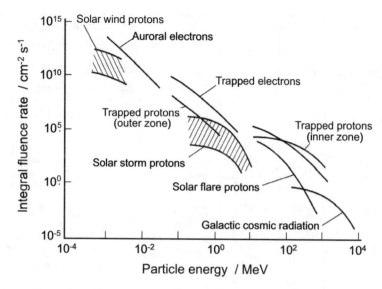

Fig. 2.1 Approximate energy ranges and spectra of different components of the radiation field in space. (From Wilson 1978)

with the magnetic field in the interplanetary medium. The strength of this effect depends primarily on the particle's energy and charge and on the activity of the Sun and the related solar wind velocity and is accordingly strongest during periods of strong solar activity (solar maximum) and weakest during low solar activity (solar minimum). This leads to a GCR intensity that is anti-correlated with solar activity showing intensity maxima at times of low solar activity and vice versa.

For the radiation exposure in space, only the hadronic part of the GCR is relevant. The intensity of electrons and positrons in the GCR is on the order of 1–2% (Boezio et al. 1999, 2000), and the dose is even lower as the dose per fluence is, depending on the particle energy, significantly lower than for protons and heavier ions. The hadronic part consists mainly of protons ($\approx$87%), alpha particles ($\approx$12%), and to a lesser extent of heavier ions (Simpson 1983). Even though the abundance of heavier ions is low, their contribution to the exposure is significant and accurate consideration in measurements and models is crucial. Model predicted GCR abundances in near-Earth interplanetary space and averaged over the ISS orbit are illustrated in Fig. 2.2.

The relevant energy range for the exposure from GCR depends on the specific shielding environment but typically ranges from approximately 100 MeV/n to 100 GeV/n (Mrigakshi et al. 2013a).

For heavily shielded environments including additional magnetic shielding energies up to 1 TeV/n can become relevant. This is the case, for instance, for low latitude flights in aviation.

In the energy spectrum above a few ten GeV up to 1 TeV, the GCR energy spectra follows a power law with a spectral index of approximately −2.7 (Blasi 2013) but a more detailed consideration shows that the index is neither constant for different

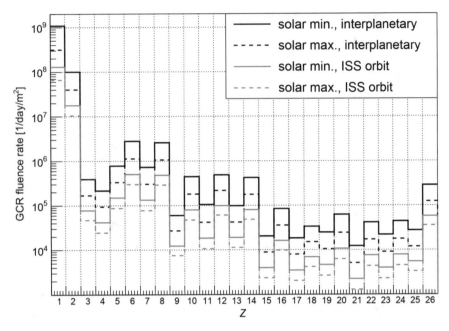

Fig. 2.2 Particle fluence rate for GCR with charge number Z = 1 (hydrogen) to 26 (iron) during solar maximum and minimum conditions in near-Earth interplanetary space and at the orbit of the International Space Station (ISS) as predicted by the model by Matthiä et al. (2013)

energies nor identical for all GCR particles (Adriani et al. 2011; Aguilar et al. 2015a, b), for instance, showing a hardening of the proton spectrum above 200 GeV (Aguilar et al. 2015b). At even higher energies above the so-called knee at around $3 \cdot 10^6$ GeV, the energy spectrum becomes steeper. At energies below a few ten GeV, the spectrum gradually bends over showing a maximum at around 500 MeV/n. This region is strongly affected by the solar modulation and varies in intensity during the solar cycle. Figure 2.3 illustrates the GCR spectra of hydrogen (proton), helium (alpha), and iron nuclei as described by the models of Matthiä et al. (2013) and Badhwar-O-Neill 2014 (NASA 2015) for solar minimum and maximum conditions during the years 2010 and 1991, respectively. Data for the Badhwar-O'Neill model have been taken from NASA's OLTARIS tool (*oltaris.nasa.gov*). The variation in the GCR flux at 100 MeV/n is about one order of magnitude and decreasing with increasing energy. Particles with energies above a few tens of GeV/n are almost unaffected by solar modulation. The total fluence rate integrated over energies between 10 MeV/n and 200 GeV/n of nuclei from H (Z = 1) and Fe (Z = 26) for solar minimum and maximum and for interplanetary space and ISS orbit is given in Fig. 2.3. It shows that hydrogen nuclei are more affected by solar modulation than heavier nuclei and that the variation in the ISS orbit is much weaker than in interplanetary space. The latter is a consequence of the magnetic shielding in the ISS orbit deflecting a significant part of the lower energetic primary particles which are most affected by the solar modulation. The total fluence in the average ISS orbit is

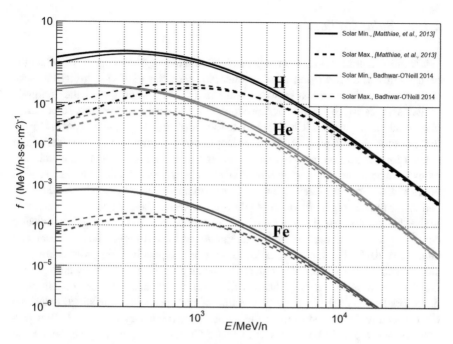

Fig. 2.3 GCR spectra of protons (H), alpha particles (He), and Fe ions for solar minimum and maximum as described by two models: Matthiä et al. (2013) and Badhwar-O-Neill 2014 (NASA 2015)

reduced by about one order of magnitude compared to interplanetary space which is a combined result of the magnetic shielding and the shielding provided by the Earth.

The variation in the GCR intensity over the solar cycle translates to changes in the radiation exposure in space, the magnitude of which depend on the shielding situation and the dose quantity. Mrigakshi et al. (2013b) have estimated that in a lightly shielded environment the variation between the most extreme GCR minimum and maximum in the past decades was expected to be about a factor of 3 for interplanetary space and about a factor of 2 for an average ISS orbit. NASA (1999) gave similar values for an unshielded interplanetary environment and predicted decreasing amplitude for increasing mass shielding. The mass shielding in interplanetary space has a similar effect as the magnetic shielding in a low inclination LEO: it reduces the relative contribution of lower energetic primary GCR which are the cause for the solar modulation-driven variation in the dose rates. It is evident that the estimation of the exposure from GCR depends on the accuracy of the underlying GCR model. Mrigakshi et al. (2013a) and Slaba and Blattnig (2014) have investigated the effect of applying different primary GCR models in the prediction of the radiation exposure. The results showed that differences in the predicted dose rates using different models for the primary GCR can easily exceed 50% if the GCR model is not sufficiently benchmarked.

2.1.2   Solar Radiation

While the Sun produces electromagnetic radiation over a wide range of wavelengths permanently, the production of solar energetic particles (SEPs) is limited to sporadic events which can last between hours and days. The origins of the energetic particles are active regions on or close to the visible surface of the Sun, solar flares, or coronal mass ejections (CMEs) close to the Sun or in interplanetary space. During these events, charged particles, mostly electrons and protons and a minor fraction of heavier ions, are accelerated to relativistic energies. Impulsive events in which particles are accelerated in magnetic reconnection events in the solar flare last several hours and are of minor relevance for radiation exposure due to their shorter duration and lower fluence compared to gradual events in which the particles are accelerated in the shock accompanying the CME. These events are longer lasting and typically contain a larger fraction of highly energetic protons (Reames 2013; Desai and Giacalone 2016). For radiation exposure purposes typically, only protons are considered and other particles are neglected. The characteristics of SEP events as they are observed at Earth vary significantly from event to event and depend on a number of factors such as the magnetic connection of the observer to the shock front, the CME speed and the conditions of the ambient interplanetary medium. Larger particle fluence typically correlates with faster CMEs speeds and events which are observed on the western hemisphere of the Sun. The time profile of the particle flux during an event depends on the energy that is considered and how the observer is connected to the source of the energetic particles (Reames 1999; Cane and Lario 2006). The rise time between the onset of the event and the peak of the particle flux can be between minutes and days and also the durations of events vary. For relatively low energies up to tens of MeV, the maximum may be reached only when the shock driven by the CME passes the observer. These are called energetic storm particles (ESP).

The energy spectrum of the SEPs reaches from the keV region up to several hundred MeV. Protons during the most intense of these events can even reach kinetic energies of several GeV and if the particle fluence is great enough, the increased intensity of cosmic radiation can also be recorded on ground by neutron monitor stations (Simpson 2000). These events are then called ground level events or enhancements (GLE). In recent decades, GLEs have occurred on average approximately once a year with an increased frequency of occurrence during periods of high solar activity. Between 2010 and 2020, however, only two such events have been recorded, on 17 May 2012 and 10 September 2017 which is a consequence of the modest solar activity in the current solar cycle. Solar particle events with lower intensity and particle energy are more frequent and Gopalswamy et al. (2015) list 37 large SEP events between Aug 2010 and Nov 2014 which is still a low number compared to previous solar cycles. The energy spectrum of the protons can be described by a single or double power law in energy or rigidity. The slope of the spectrum is of great importance for the impact of the event on radiation exposure. Soft spectra with a large fraction of lower energetic particles are more easily

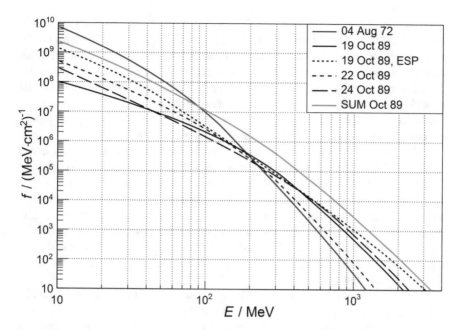

Fig. 2.4 Energy spectra of selected large SEP events as described by Tylka et al. (2010)

shielded by mass or by a magnetic field and the radiation exposure in the human body is expected to be less homogeneous than in case of a hard event which contains a relatively larger fraction of highly energetic particles.

Figure 2.4 illustrates the event integrated differential fluence spectra derived from Tylka et al. (2010) who described the integral proton spectrum by a double power law in rigidity with an exponential turnover (Band et al. 1993). The sum of the October 1989 events was recommended to NASA by Townsend et al. (2018) to be used as standard in designing storm shelters against solar particle events space habitats.

The October 1989 event had a comparatively hard spectrum meaning that it contained a large fraction of highly energetic protons with energies above a few hundred MeV. These particles can penetrate moderate or even heavy shielding (the range of 200 MeV protons in Al is approximately 120 mm and of 1 GeV protons approximately 1.5 m) and contribute to the exposure directly but with reduced intensity. Townsend et al. (2018) estimated that organ dose rate for the combined Oct 1989 event would exceed the exposure limit of 250 milliGray-Equivalent (mGy-Eq, see 2.2.1) for shielding below approximately 10 g/cm² Al (≈4 cm). Softer events may have greater fluence at lower energies which could lead to higher doses for lightly shielded conditions, for instance, during an EVA. The total dose that is to be expected is always a combination of the particle spectra, the temporal profile, and the exposure time under the specific shielding conditions. For instance, the dose from a soft event could be greater for a lightly shielded environment than the dose from a hard event if the fluence at low energies is higher. For the identical events encountered at heavier shielded locations, on the other hand, the situation may be inverted and the greater number of highly energetic particles in the spectrum of the hard event may lead to the higher doses. The total dose to which an astronaut would be exposed to during an event is a complex combination of the event characteristics, for instance, onset to peak rise time, spectral hardness, and total duration, and the specific exposure conditions like shielding environment and duration of stay at different locations. During large events with a short rise time and a sharply peaked maximum, for instance, it is more important for the astronaut to reach a better shielded location (a radiation shelter) quickly than for an event which may have a similar total fluence but a longer rise time with a less sharply peaked maximum.

General statements on the impact of SEP events on the dose are therefore hard to make but most numerical estimates of the expected dose rates agree that for interplanetary space a mass shielding equivalent to approximately $10–20$ g/cm^2 is sufficient to reduce the exposure below current thresholds (Townsend et al. 2018). For other exposure scenarios, like LEO, Moon, or Mars surface, the required mass is reduced due to the additional shielding provided by the magnetosphere, the atmosphere, and the planet.

As the occurrence frequency of SEP events varies over the solar cycle and also from cycle to cycle, probabilistic models can be used to estimate the risk for astronauts to be exposed to an event of a specific magnitude for a certain mission scenario depending on the size of the event, the length of the mission, and the launch date relative to the solar cycle, e.g., Xapsos et al. (2000), Kim et al. (2009), Jiggens et al. (2018).

2.1.3 Radiation Belt Particles

The term radiation belt refers to regions surrounding Earth, or more generally any planet with a global magnetic field, and containing highly energetic charged particles, mainly protons and electrons. These energetic particles are temporarily trapped in the magnetosphere in toroidal structures that stretch, in case of Earth, from a few hundreds of kilometers above ground out to several ten thousands of kilometers. The radiation belts are dynamical structures which are affected by space weather, for instance, by the solar wind pressure, heliospheric magnetic field, and GCR intensity (Baker et al. 2018). Typically, Earth is surrounded by two radiation belts, an inner and an outer belt (Fig. 2.5). The inner belt contains mostly protons with energies up to several GeV and a small fraction of electrons. The dominant source of high energy protons in the inner radiation belt is the cosmic ray albedo neutron decay (CRAND) process (Jentsch 1981) but other sources can contribute.

Depending on altitude and inclination, objects in an Earth orbit may cross the radiation belts several times per day and may be exposed to varying intensities of radiation belt particles with a wide range of energies. Of special interest for human

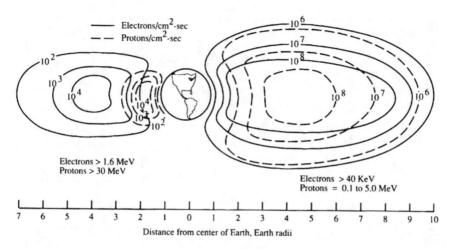

Fig. 2.5 Particle fluence rates of electrons and protons in the inner and outer radiation belts. (From NASA 1991)

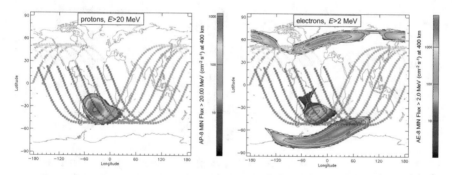

Fig. 2.6 Trapped protons (left) and electrons (right) in the radiation belts as predicted by the AP8 model in SPENVIS (www.spenvis.oma.be) at an altitude of 400 km, the approximate altitude of the ISS. The overlays show several passes of an orbit of 51.6° inclination

spaceflight is an area above the south-eastern part of South America and the South Atlantic, the so-called South Atlantic Anomaly (SAA). In this area, the inner radiation belt approaches the surface of Earth down to a few hundreds of kilometers above ground, due to the tilt and shift of the axis of the dipole-like magnetic field of Earth with respect to its axis of rotation. As a consequence, objects in a LEO, in particular the ISS, cross the SAA several times per day and are exposed to the significantly increased particle flux within this region (Fig. 2.6).

Figure 2.7 shows a ten-day average during solar minimum of the differential energy spectra in an ISS like 51.6° inclined orbit at an altitude of 400 km derived from AP8/AE8 (Vette 1991) and AP9/AE9 (Ginet et al. 2013) models in SPENVIS

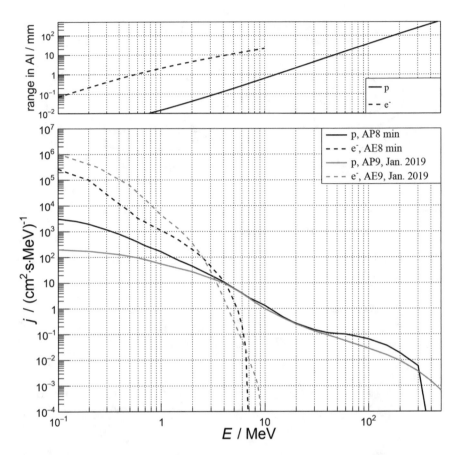

Fig. 2.7 Differential proton (p) and electron (e⁻) spectra from AP/AE8 and AP/AE9 models for a 51.6° inclined orbit at an altitude of 400 km (bottom) and the corresponding particle ranges in aluminum (top)

(www.spenvis.oma.be). The respective range of protons and electrons in aluminum is illustrated in the top panel. The trapped particle environment encountered in this orbit, according to the models, contains protons with energies up to several hundred MeV from the inner radiation belt in the SAA (cf. Fig. 2.6) and electrons with energies up to almost 10 MeV mostly from the outer radiation belt.

Electrons below a few MeV are stopped by a few millimeters of aluminum while radiation belt protons at their highest energies can penetrate several centimeters of shielding. In combination with the fact that the electron intensity drops extremely fast by several orders of magnitude between 1 and 10 MeV means that an increase in dose from electrons is not observable at the ISS orbit inside the station while the contribution of the trapped protons to the absorbed dose can reach 50% and more.

2.2   Radiation Dosimetry

2.2.1   Dose Quantities

A framework of dosimetric quantities, definitions, and recommendations concerning radiation protection has been established and continuously developed by the International Commission on Radiological Units and Measurements (ICRU) and the International Commission on Radiological Protection (ICRP). This chapter summarizes the most important quantities relevant for space exploration: relevant reports are: "The 2007 Recommendations of the International Commission on Radiological Protection" (ICRP 2007), "Assessment of Radiation Exposure of Astronauts in Space" (ICRP 2013), "Adult Reference Computational Phantoms" (ICRP 2009), and "Fundamental Quantities and Units for Ionizing Radiation (Revised)" (ICRU 2011). Here, only the most relevant quantities are introduced and the interested reader is referred to the ICRU and ICRP reports for further detail.

On Earth, radiation protection is based on effective dose and ambient dose equivalent, for instance, for the exposure to cosmic radiation in aviation (ICRP 2016). In space, however, effective dose is not applicable due to the use of a single radiation weighting factor $w_r = 20$ for heavy ions which is not generally a valid choice for highly energetic nuclei from GCR (ICRP 2013). Instead, the use of effective dose equivalent is recommended which is based on the quality factor Q rather than radiation weighting factors. No operational dose quantity has been found to be applicable to the exposure to cosmic radiation in space by the ICRP.

The basic quantity in dosimetry on which many of the derived quantities in radiation protection are based on, is the absorbed dose D which is defined through ε, the energy imparted in a volume of mass m. Absorbed dose is defined as:

$$D = \frac{d\varepsilon}{dm} \tag{2.1}$$

The unit of the absorbed dose is J/kg $\equiv$ Gy (gray).

The absorbed dose in macroscopic volumes can be derived by integrating over the volume of interest. If the target volume is a certain type of tissue or organ T in the human body, the absorbed dose in this volume is denoted D_T.

To consider the biological relevance of different types of radiation, the quality factor Q is introduced and the dose equivalent H is defined as:

$$H = Q \cdot D \tag{2.2}$$

The unit of the absorbed dose is J/kg $\equiv$ Sv (sievert).

Equivalently, if the absorbed dose in a tissue T is considered, the corresponding dose equivalent is calculated as $H_T = \overline{Q}D_T$, where $\overline{Q}$ is the mean quality factor in the tissue. The quality factor as defined by (ICRU 1986; ICRP 1991) is a dimensionless factor, solely depending on the linear energy transfer of the particle depositing the energy. The linear energy transfer L_Δ or equivalently, the restricted linear electronic stopping power, is defined in ICRU (2011) as "the quotient of dE_Δ by dl, where dE_Δ is the mean energy lost by the charged particles due to electronic interactions in traversing a distance dl, minus the mean sum of the kinetic energies in excess of D of all the electrons released by the charged particles":

$$L_\Delta = \frac{dE_\Delta}{dl} \tag{2.3}$$

The unrestricted linear energy transfer $L_\infty \equiv L$ is identical to the electronic stopping power. Based on this definition, the quality factor Q has been defined as a continuous function of L (or LET) in water in ICRP 1991:

$$Q(L) = \{ \begin{array}{l} 1, \text{if } L < 10\,\text{keV}/\mu m \\ 0.32L - 2.2, \text{if } 10 \le L \le 100\,\text{keV}/\mu m \\ 300/\sqrt{L}, \text{if } L > 100\,\text{keV}/\mu m \end{array} \tag{2.4}$$

The quality factor is unity for all low L particles, such as electrons and positrons, mostly one for muons, pions, and can become larger for protons and heavier nuclei. The maximum quality factor of 30 is reached at 100 keV/μm.

The quality factor at a point in tissue is defined in ICRP (2013) as

$$Q = \frac{1}{D} \int_\infty^{L=0} Q(L) D_L dL \tag{2.5}$$

where D is the absorbed dose in tissue and $DL = dD/dL$ is the distribution of D in L (for charged particles in water) at the point of interest.

Based on the dose equivalent, the ICRP defines the whole body quantity effective dose equivalent H_E which is a weighted sum over the organ dose equivalents using the tissue weighting factors w_T (Eq. 2.7). Other than in terrestrial radiation protection for which the use of the effective dose is recommended, ICRP (2013) recommends to use the effective dose equivalent in the cosmic radiation field in space.

The most recent tissue weighting factors (Table 2.1) introduced in ICRP (2007) are age and sex averaged; the weighting factors are based on experimental data covering stochastic effects (radiation-induced cancer and heritable diseases).

The effective dose equivalent is defined as:

$$H_E = \sum_T w_T H_{T,Q} = \sum_T w_T Q_T D_T \tag{2.6}$$

where the sum is over all tissues T listed in Table 2.1 and Q_T and D_T are the organ averaged quality factor and the absorbed dose in the respective tissue.

For a known radiation field, ICRP (2013) provides fluence-to-dose conversion coefficients for relevant particles, organs, and the total effective dose equivalent for

Table 2.1 Tissue weighting factors w_T recommended by the ICRP

Tissue	w_T	Σw_T
Bone marrow (red), colon, lung, stomach, breast, remainder tissues[a]	0.12	0.72
Gonads	0.08	0.08
Bladder, esophagus, liver, thyroid	0.04	0.16
Bone surface, brain, salivary glands, skin	0.01	0.04
	Total	1.00

From ICRP (2009)

[a]Remainder tissues: Adrenals, Extrathoracic (ET) region, Gall bladder, Heart, Kidneys, Lymphatic nodes, Muscle, Oral mucosa, Pancreas, Prostate ($\male$), Small intestine, Spleen, Thymus, Uterus/cervix ($\female$)

a reference anthropomorphic phantom (ICRP 2009) which can be folded with the particle fluence spectra to calculate the dose in the field.

The NASA effective dose as defined in NASA (2013) is not to be confused with the effective dose defined by the ICRP. It is formally identical, however, with the effective dose equivalent concerning the use of quality and tissue weighting factors but with differing numerical values. As stated above, in the ICRP effective dose, the absorbed dose is weighted with radiation weighting factors. NASA applies a risk model that separates the risk of fatal solid cancer and leukemia and the applied quality factors not only depend on the linear energy transfer L but also on the parameter $Z*^2/\beta^2$ with $Z*$ and β being the effective charge number and the particle velocity relative to the speed of light, respectively (Chap. 4). Additionally, the ICRP includes the risk of non-lethal cancer into its gender and age averaged tissue weighting factor while NASAs Q is based on cancer mortality risk.

For the non-stochastic (deterministic) radiation risk, both NCRP (2000) and (ICRP 2013) recommend using the absorbed dose in an organ (D_T) weighted with the relative biological effectiveness (RBE, Sect. 3.3), the gray-equivalent (Gy-Eq). Values for the RBE recommended by ICRP and NCRP to be used in the context of exposure to cosmic radiation are between 1 and 6.

Based on the above introduced quantities, space agencies develop their radiation protection framework and dose limits. For stochastic radiation effects, the Russian, European, and Canadian Space Agencies use the ICRP recommended career limit of 1 Sv. JAXA and NASA, on the other hand, use limits on the Risk of Exposure-Induced Death (REID) which lead to age- and gender-specific limits in the dose. In addition to limits related to stochastic effects, the agencies also introduce dose limits on different organs to consider non-cancer effects; for NASA astronauts, for instance, dose limits for 30 days, one year and the whole career for non-cancer effects are in place (NASA 2014). ESA defines annual limits and for 30 day periods for blood-forming organs, eye and skin (Straube et al. 2010). Current dose limits are summarized in McKenna-Lawlor et al. (2014).

2.2.2 *Radiation Detectors and Their Calibration*

Dosimetry requires a dedicated radiation detector able to determine the relevant quantities of the radiation field under study. For applications in space radiation dosimetry, the detector system has to be able to measure the relevant radiation protection quantities as defined in the previous chapter, except for the effective dose equivalent which can only be calculated. This implies, that one has to apply detector system being able to measure in what way ever the absorbed dose, the linear energy transfer (LET) spectra and thereby the quality factor of the radiation field as well as the dose equivalent. The effective dose equivalent, on the other hand, would further on be a quantity which can only be measured in space by applying relevant anthropomorphic phantoms.

For the measurement of the space radiation field, one can distinguish between two detector principles. The first detector principle is represented by passive radiation detectors, being able to store the relevant energy deposition from ionizing radiation in their detection material. Examples of passive radiation detectors are thermoluminescence (TL) or optical stimulated luminescence (OSL) detectors as well as Nuclear Track Etch Detectors (CR-39). The second type of instruments are active (in some way powered) instruments which are, for example, based on silicon detectors or on the principle of tissue equivalent proportional counters. Both systems applied have their advantages and disadvantages. While the passive systems have low mass and small dimensions, do not need any external power and data interface and can easily be placed at various positions inside a spacecraft, they do not offer time-resolved data, and they usually have to be returned to the laboratory for evaluation. At the end, they will provide one data value integrated over their respective exposure time.

In contrast to this, the active detector systems enable the investigators to have time-resolved data thereby also having the possibility to resolve the changes of the radiation environment on short time scales—as for example during an SPE. It has to be nevertheless taken into account that the active systems need to be provided with a power and data interface, which can be a demanding task for certain applications. For decades, both of these detector principles have been applied on various space missions and now especially in the frame of the upcoming planned exploration missions to the Moon and in the future to Mars the active detectors will become the main instruments to be applied offering real-time data capabilities as well as possible alarm capabilities for extreme radiation events, as for example an SPE. A relevant summary of instruments applied on-board the ISS as well as for mission to the Moon and Mars is provided in (Berger 2008; Caffrey and Hamby 2011; Narici et al. 2015).

All instruments have in common that they need to be calibrated to the relevant components of the space radiation field. This can be and is accomplished at various facilities around the world offering reference radiation fields for the respective particle species. As for example monoenergetic neutrons are provided at PTB,

Braunschweig, Germany, while the CERF neutron reference field at CERN, Switzerland, provides a neutron spectrum similar to high altitudes.

Of special importance for space are calibrations at facilities offering heavy ions to simulate the GCR environment encountered in space. One of these facilities is the Heavy Ion Medical Accelerator (HIMAC) at the National Institute of Radiological Sciences in Chiba, Japan (Fig. 2.8). Applying the HIMAC facility the space radiation community started in the early 2000 the ICCHIBAN project, aiming for a comparison of the properties of passive and active radiation detectors, thereby also enabling to build up a database of relevant instrument properties (Uchihori et al. 2002; Yasuda et al. 2006). In the last 20 years, a lot of effort has been put in the calibration and comparison of various passive (e.g., Berger and Hajek 2008) and active (Berger et al. 2019) radiations detector systems applied in space and almost all of the

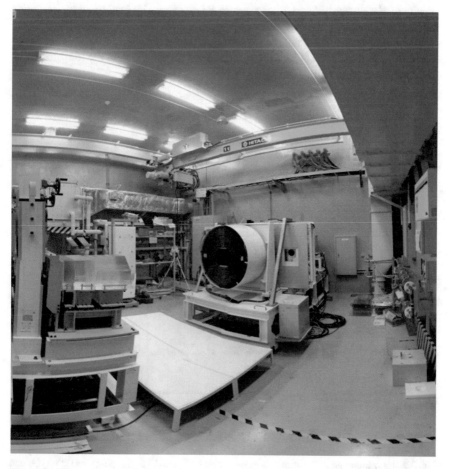

Fig. 2.8 Heavy Ion Medical Accelerator (HIMAC) at NIRS, Chiba, Japan. (© Bartos Przybyla, DLR)

active instruments currently applied for radiation measurements on-board the ISS, the Moon or Mars have been calibrated at the HIMAC facility (Benton et al. 2019).

2.2.3 The History of Space Radiation Dosimetry

In 1912, Victor Franz Hess (1912) started his famous balloon flight, which led to the discovery of the galactic cosmic radiation, and resulting in bestowal of the Nobel Prize to V. Hess in 1936. Though at the time of discovery Hess called the radiation, he discovered "Höhenstrahlung" stating (in the original German written manuscript): "...*Die Ergebnisse der vorliegenden Beobachtungen scheinen am ehesten durch die Annahme erklärt werden zu können, daß eine Strahlung von sehr hoher Durchdringungskraft von oben her in unsere Atmosphäre eindringt...*" ("*... The results of the present observation seem to be most readily explained by assuming that radiation of very high penetrating power enters the atmosphere from above, and can still produce a part of the ionization observed in closed vessels at the lower altitudes*"). Figure 2.9 provides the results of his flights showing the ionization rate in dependence on the altitude above sea level.

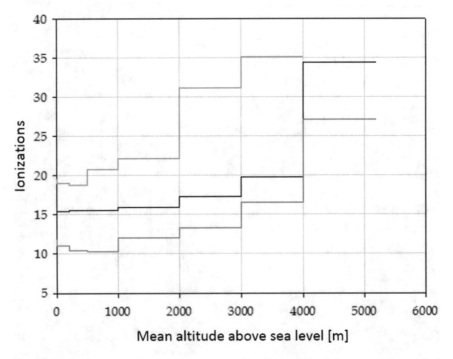

Fig. 2.9 Ionization rate in the atmosphere measured by V. Hess. (Graph drawn based on original data given in Hess 1912)

Further investigations in the coming decades lead to the discovery that these particles are charged ions and that they originate from outside our solar system, which lead in the end to the name cosmic radiation, and to a further Nobel Prize for Cecil Powell in 1950 (Powell 1950). A further historical overview of these endeavors is provided in Carlson (2012). The late 1950s with the International Geophysical Year (IGY) in 1957–1958 led to the discovery of the Earth's radiation belts (Ludwig 1962).

These were discovered by placing various radiation detectors as for example on the US Explorer satellites, which at the end led to the naming of the radiation belts after the main US scientist James Van Allen (Van Allen and Frank 1959; Van Allen et al. 1959a, b). One shall not forget that also Sergei Vernon (Vernov and Chudakov 1960) from the Moscow State University placed a Geiger-Müller counter on the Sputnik 2 mission and also came to the same conclusion as James Van Allen (Baker and Panasyuk 2017). So at the beginning of the human space age in the early 1960s, it was clear that Earth was surrounded by radiation belts and also that cosmic radiation impinges on the Earth atmosphere.

Nevertheless, the reason for starting radiation measurements for human space missions in the USA was a different one. The amount of nuclear explosions in the atmosphere (Hess 1964) and the subsequent creation of artificial radiation belts especially after the "Starfish Prime" nuclear explosion in 1962 shortly before the Mercury mission led to the statement *"...the creation of an artificially trapped electron belt by a high-altitude nuclear explosion on July 9, 1962, made it necessary to place radiation dosimeters aboard the spacecraft used in the eight Mercury-Atlas mission (MA-8)..."* (Warren and Gill 1964). This was the starting point of the radiation measurements for human spaceflight. Figure 2.10 shows how the astronauts on the Mercury mission were equipped with passive thermoluminescence detectors on various parts of the body.

From the NASA Mercury mission onwards (Warren and Gill 1964) followed by the Gemini missions (Richmond 1972), all astronauts were equipped with personal passive radiation detectors and additional instruments as nuclear emulsion and ionization chambers were applied within the spacecraft. During the Apollo missions, various experiments were carried out to determine the radiation loads as given in English et al. (1973), Schaefer et al. (1972), and Schaefer and Sullivan (1976).

An overview table of the radiation doses received during the Apollo lunar landing missions is given in Table 2.2, based on the compendium of space-related dosimetry data provided by Benton (1984).

Noteworthy at this point is also that the first German experiment, the Biostack experiment (Buecker et al. 1973) was already flown as part of the Apollo 16 and 17 missions with its aim to correlate the radiation environment outside LEO and its effect on biological samples. The total mission dose measured with this experiment by thermoluminescence detectors was between 5.0 and 6.2 mGy, which is very close to the data provided by the NASA detectors (as given in Table 2.2 for the Apollo 16 mission).

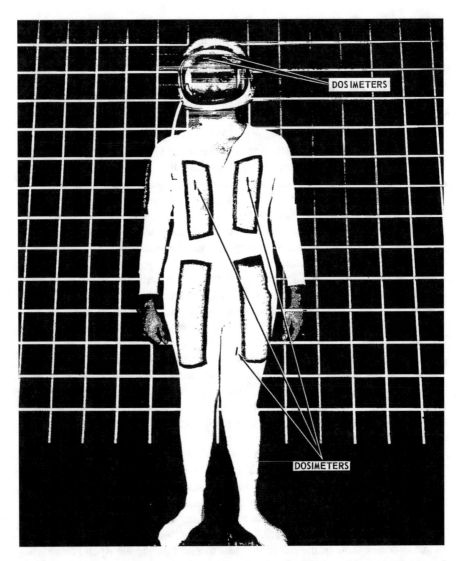

Fig. 2.10 Placement of TLD detectors on the pilot for the Mercury mission. (Warren and Gill 1964)

One should not forget when thinking about the space race to the Moon that also the scientists in Russia were working on radiation detectors and radiation dosimetry for the Russian spacecraft. An overview of data measured during the Vostok, Voshkod (1961 to 1965), and the Soyuz-3 to −9 missions in the years 1968 to 1970 is provided in Benton and Henke (1983). With the start of the NASA Space Shuttle missions, a new spacecraft was available for science, but also as a vehicle for space dosimetry-relevant first data from the STS-1 to the STS-41 missions are

Table 2.2 Overview of Apollo missions

Apollo mission	Duration (hours)	Lunar surface duration	Average radiation dose [mGy]
11	194.0	21 hrs, 38 mins	1.8
12	244.5	31 hrs, 31 mins	5.8
14	216.0	33 hrs, 31 mins	11.4
15	295.0	66 hrs, 54 mins	3.0
16	265.8	71 hrs, 2 mins	5.1
17	301.8	74 hrs, 59 mins	5.5

Data from Benton (1984)

summarized in Benton (1984). The following MIR space station already had radiation instrument (as for example the famous R-16 ionization chamber) installed inside to provide baseline data and possible alarm capabilities for the crew. With the increased cooperation in space, seen for example by the visits of the Space Shuttle to MIR (so-called Shuttle-MIR missions), the amount of groups providing detectors and comparing their results increased. Badhwar et al. (2002) provide a good compendium of all instruments applied within the MIR space station, their properties and provide also an overview and comparison of measured and calculated radiation values for various locations inside the MIR space station. A full comparison of all measurements performed on-board MIR by various institutions is given in the special issue "Radiation on the MIR Space Station" (Radiat Meas. 35, 5, 2002).

With the launch of the ISS at the end of the 1990s radiation detectors were an integrated part of the station in the USA and the Russian segment and were and still are applied as operational detectors together with additional instruments aiming for various scientific radiation research as given in Berger (2008) and Narici et al. (2015).

2.2.4 Human Phantom Experiments

One of the foremost objectives of space radiation dosimetry is to serve as a solid basis for risk assessment for cosmic ray-induced late effects such as cancer, even if the radiobiology associated with these unique radiation fields to a considerable amount is still unknown. The use of dedicated phantoms, simulating a space traveler's body, provides a detailed mapping of dose distribution that is essential for evaluating the doses absorbed in different organs and tissues. Dose measurements are obtained by using a generally large amount of miniature dosimeters, such as thermoluminescence (TL) phosphors, arranged in a regular grid within the mannequin. Due to the considerable mass of such phantom bodies, the number of associated experiments that have actually been conducted in LEO is small. Table 2.3 lists the phantom experiments performed during various space missions.

The experiments started with the first anthropomorphic phantom head on-board a Space Shuttle in the years 1989–1990 (Konradi et al. 1992), followed by the

Table 2.3 Phantom experiments in Space

Experiment	Date	Location
Phantom Head	1989–1990	STS-28 / STS-36 / STS-31
Spherical Phantom	1997–1999	MIR
Anthropomorphic Phantom FRED	1998	STS
Anthropomorphic Phantom FRED	2001	ISS
Spherical Phantom MATROSHKA-R	2004–	ISS
Anthropomorphic Phantom MATROSHKA	2004–2012	ISS

spherical phantom on-board a space station (MIR) in the years 1997–1999 (Berger et al. 2004) and the first exposure of a whole anthropomorphic upper torso in space (Space Shuttle 1998). This space shuttle flight was the first flight, where the effective dose equivalent for a human was determined based on data from radiation detectors placed inside a phantom (Yasuda 2009). This torso (so-called FRED) was also applied for the first torso measurements inside the ISS during Increment 2 in the year 2001. A second—higher developed—spherical phantom (MATROSHKA-R) started its measurement phase inside the ISS in the year 2004 (Kireeva et al. 2007), together with the ESA MATROSHKA (MTR) experiment (outside the ISS in the MTR-I phase) (Reitz et al. 2009). The MATROSHKA experiment was further on also performed inside the Russian part of the ISS in the frame of the MTR-2A and -2B phases and also in the Japanese part of the ISS in the frame of the MTR-2 KIBO experiment. This was the first long-term exposure of a phantom at one hand outside the ISS and on the other hand for three long missions inside the ISS.

As stated before, the ESA MATROSHKA facility and the respective MATROSHKA experiments were the biggest endeavors for the determination of the effective dose equivalent ever accomplished on-board the ISS.

Within the MTR facility thousands of passive thermoluminescence detectors were applied to determine as close as possible the relevant organ doses inside the phantom. In addition, detectors were placed on the surface of the phantom to determine the skin dose (Berger et al. 2013) and the dose equivalent on the surface to have a comparison with the finally determined effective dose equivalent (Puchalska et al. 2014). For more information about the MATROSHKA experiment—see also https://www.fp7-hamlet.eu. Figure 2.11 provides on the left a picture of the MTR facility mounted outside the Russian part of the station in the frame of the MTR-1 experiment and on the right the results for the three-dimensional dose distributions measured during this exposure. These results are the baseline for the calculation of the organ doses and for the further determination of the effective dose equivalent.

In summary it was evaluated that the effective dose equivalent (as given in Table 2.4) for an outside exposure (MTR-1) is a factor of two to three lower than the measured dose on the surface of the body (skin dose) due to the self-shielding of the body for the lower energetic electrons and protons encountered outside the ISS. For an inside exposure, a personal dosimeter would still conservatively overestimate the effective dose equivalent to approximately 20%.

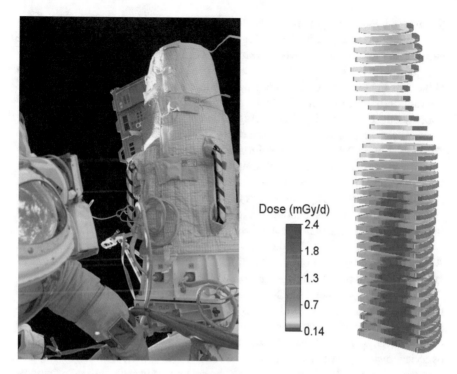

Fig. 2.11 The MTR-1 outside exposure and the 3D dose distribution from the outside exposure. (Data DLR)

Experiment	E (μSv/d)	Skin dose (μSv/d)
MTR-1 (outside ISS)	722 ± 35	3025 ± 453
MTR-2A (inside ISS)	552 ± 26	641 ± 96

Table 2.4 Effective dose equivalent (E) for two MTR experiments

From Puchalska et al. (2014)

2.3   Exposure Scenarios: Measuring and Modeling of Space Radiation

If the radiation exposure in different space scenarios is considered for instance by comparing different measurements or model calculations, it is of paramount importance to keep in mind that the environment is extremely variable in time and space but also strongly depends on the specifics of the dose measurement or calculation. The introduction of a small detector into a radiation field may lead to negligible changes in the field in most cases but the human body, may it be in simulations by means of numerical phantoms, in measurements by means of an anthropomorphic or water phantom or in personal dosimetry, will affect the radiation field and, as a consequence, the dose rates. This effect will be most pronounced in fields which are dominated by low energy particles, such as most SEP events and outer radiation belt electrons, but is also present for the highly penetrating field of GCR. In an extreme case a small detector may measure extremely high dose values from relatively low energetic particles which would, in case of human exposure, be completely absorbed in the skin and the dose to the more sensitive inner organs may be zero. Additionally, the variation of the radiation field with shielding is not necessarily linear and deriving organ dose values using their average shielding is not always a valid approach.

Many detectors that are applied in dose rate measurements use silicon chips as sensitive volume. The energy deposition of neutrons in silicon, however, is very low compared to water or tissue, the relevant materials for radiation protection. In a situation in which a relevant secondary neutron field exists, the additional contribution from neutrons to the dose in tissue has to be evaluated by other means.

It follows that extreme caution has to be taken if results that have been obtained in one scenario are translated to a different one, for instance, if measurements from a sub-millimeter silicon detector are used to derive organ doses to humans. Additional information about the radiation field, for instance, through model calculations or other types of detectors, is absolutely necessary in such a case.

The following chapter describes different scenarios that are of importance to current human spaceflight or will become important in the near future. In all of these scenarios, astronauts are constantly exposed to GCR and potentially to SEP and, in case of LEO, to charged particles in the radiation belt. Figure 2.12 gives a rough overview over approximate dose rates from GCR that are encountered in different exposure scenarios. It is important to note, however, that in each of these situations, the dose rate can vary significantly, depending for instance on altitude, location, mass shielding, or solar activity.

While there is comprehensive experimental data from the ISS which were recorded under human spaceflight conditions, no such data exist for interplanetary

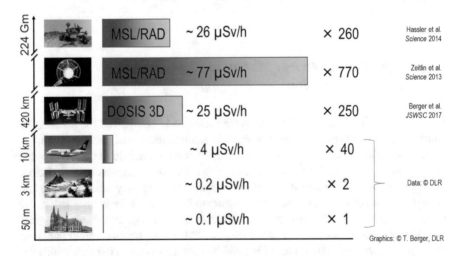

Fig. 2.12 Approximate dose rates in different exposure scenarios

space and the lunar and Martian surface. Available information for these scenarios is restricted to measurements on robotic missions and model estimates.

2.3.1 Low Earth Orbit: The International Space Station

The International Space Station (ISS) is in a LEO at an altitude between approximately 300 km and 400 km and with an inclination of 51.6° which means that it reaches maximum geographic latitudes of 51.6°N and 51.6°S, respectively. Its orbiting period is approximately 90–93 minutes which corresponds to 15.5 to 16 orbits per day. As a consequence, the longitude of the ascending node of the ISS shifts west by approximately 23° for each pass.

The radiation environment to which the ISS and the astronauts are exposed to varies significantly within one orbit but also during one day, when the longitude of the ascending node changes. The underlying mechanism of these variations is the magnetic field of Earth and its effect on charged particles arriving at Earth from interplanetary space and the fact that charged particle populations are trapped within this field. The total exposure of astronauts on the ISS is typically about 200–300 μGy/d and 500–700 μSv/d (Berger et al. 2017); the exact value depends on the local shielding, the point in the solar cycle, the altitude of the station, and other factors.

Astronauts on-board the ISS are protected from GCR and SEPs by two natural mechanisms: The obstruction of the sky by the solid Earth and the shielding provided by the Earth's magnetosphere.

The former can be estimated as follows: If Earth or any other celestial body is approximated by a sphere with radius R, the fraction f of the obstructed sky for an object in an orbit around the body can be expressed as a function of the altitude h of the object above the surface:

$$f = 0.5 \cdot \left(1 - \cos\left(\arcsin\left(\frac{R}{h+R}\right)\right)\right) \tag{2.7}$$

For a low Earth orbit at $h = 400$ km above ground and a radius between 6357 km and 6378 km (https://nssdc.gsfc.nasa.gov/planetary/factsheet/earthfact.html) this results in $f = 0.331$ (33%) of sky which is blocked, which means that for zero magnetic shielding at high latitudes, the dose rate is expected to be one third lower than in interplanetary space for an identical shielding, if albedo particles from the atmosphere are neglected.

While the magnetic shielding has a negligible effect at high latitudes, it has a significant influence on the radiation field encountered on-board the ISS if the whole orbit is considered. Figure 2.13 shows the effective vertical cut-off rigidity R_C which is a measure of the magnetic shielding effect against charged particles from interplanetary space. The cut-off rigidity can be used as a lower threshold for the rigidity

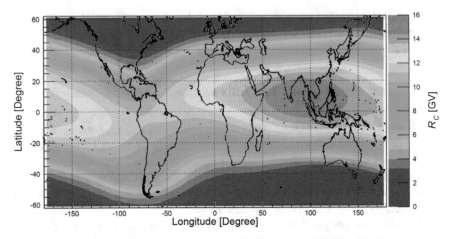

Fig. 2.13 Effective vertical cut-off rigidity R_C at 400 km altitude

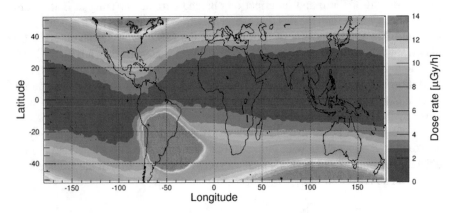

Fig. 2.14 Absorbed dose rates in silicon (Si) measured during solar minimum conditions in 2009 on-board the ISS with the DOSTEL instrument in the DOSIS project. (Berger et al. 2017)

of particles to be able to penetrate the magnetosphere to the given location. The rigidity is defined as the momentum of a particle divided by its charge $R = p/q$.

Additionally, the maximum values at the highest latitudes change during the course of the day and reach their peak values (~15 µGy/h in Si in Columbus during solar minimum) at longitudes around 100°W on the southern hemisphere and around −80°E on the northern hemisphere. As geomagnetic shielding at these positions is negligible for GCR, these values can give an estimate for the dose rates which can be expected in interplanetary space if the geometrical shielding of Earth is considered.

Figure 2.14 (Berger et al. 2017) illustrates the measured dose rates in silicon in the COLUMBUS module of ISS in 2009. The GCR flux in the year 2009 reached an intensity maximum unprecedented in the space age, and it can be considered as a worst case scenario to our current knowledge. The measured dose rate presented in Fig. 2.14 and its variation, with the exception of the South Atlantic region, are dominated by the GCR and their variation due to the magnetospheric shielding. The shielding at high south-eastern and high north-western latitudes is negligible (cf. Fig. 2.13) and the dose rates from GCR reach their peak values of more than 10 μGy/h which corresponds to about 70% of the value in interplanetary space (due to the obstruction by solid Earth at an altitude of 400 km). At low latitudes along the geomagnetic equator, the magnetic shielding is maximum and the dose rates decrease to below 2 μGy/h. The averaged dose rate from GCR in silicon in 2009 within the COLUMBUS module measured by Berger et al. (2017) was ~160 μGy/d (520 μSv/d, Q ≈ 3.3). Lishnevskii et al. (2012) determined the dose rate from GCR at different locations in the Russian service module to be between 100 μGy/d and 110 μGy/d in 2009. These values were reached during the solar activity minimum which corresponds to the GCR intensity maximum. During solar maximum, dose rates are significantly lower.

Peak dose rates in the SAA can reach values of several hundred μGy/h but the exact values depend on the trajectory of the station, especially its altitude, the local mass shielding and the solar activity, and they can change during geomagnetic disturbances caused by complex interactions of the interplanetary medium with the magnetosphere. In the relatively heavily shielded environment inside the station, the averaged dose rate from SAA particles in COLUMBUS amounts to approximately 70–100 μGy/d (100–200 μSv/d).

Inside the ISS, the astronauts are effectively shielded from the electrons in the radiation belts which reach maximum energies of a few MeV.

Although measurable in rare events, their contribution to the total dose is negligible. Outside the station, for instance during extra vehicular activities, this changes fundamentally. Dachev (2018) have measured outside the ISS between October 2014 and January 2015 an average dose rate in silicon from outer radiation belt electrons of 278 μGy/d and a maximum value of almost 3 mGy/d. As the electrons have relatively low energies, the dose rate to humans cannot directly be derived from the dose in a thin detector due to the self-shielding of the body.

Sato et al. (2011), however, have estimated by numerical means the contribution of trapped electrons to the dose to astronauts as 0.737 mSv/d for the skin and 0.0232 mSv/d for the effective dose equivalent (note: as the quality factor of electrons is unity, these numbers are identical for the absorbed dose in tissue in mGy/d). Inside the station, the estimates by Sato et al. (2011) are <1 μSv/d for both the effective dose equivalent and the skin dose.

The GCR contributions to the organ doses on-board the ISS during solar minimum for a spherical Al shielding of 20 g/cm^2 have been estimated by Matthiä et al. (2013) to be ~90 μGy/d and 220–260 μSv/d, where the lower values are for the inner organs which benefit more from the self-shielding of the body and for which the quality factor is lower.

Astronauts on-board the ISS are most of the time effectively protected from SEPs by the Earth's magnetic field. Only few events accelerate protons to kinetic energies above 1 GeV and even those can reach the ISS only at very limited regions at high latitudes at eastern longitudes in the south and at western longitudes in the north. A kinetic energy of 1 GeV corresponds to magnetic rigidity of approximately 1.7 GV for protons which limits the regions accessible to these particles to the purple areas in Fig. 2.13. As a consequence, the effects of SPEs are measurable on the ISS only during short time intervals during an orbit, if at all. Additionally, the onset of an event as measured on-board the ISS can be delayed significantly with respect to the arrival of the energetic particles at Earth due to the fact that it can take hours for the ISS to reach the areas of minimum magnetic shielding. For the most recent ground level enhancement in 2017, the delay between the beginning of the event as measured by satellites in geostationary orbit (GOES) and on the ISS was approximately 12 h (Berger et al. 2018; Matthiä et al. 2018).

Lifting space stations to higher altitudes, for instance, to reduce the atmospheric drag, results in a significantly higher radiation exposure of the astronauts as the area of and the intensity of the particle flux in the crossings of the radiation belt increases. Berger et al. (2017) have measured an increase of the absorbed dose in the SAA in the COLUMBUS module of ISS of almost 100% when the station was lifted to more than 400 km altitude from its earlier 350 km between 2011 and 2013.

Organ doses to astronauts on-board the ISS have been determined experimentally in the MATROSHKA project (Reitz et al. 2009). In the MATROSHKA project, an anthropomorphic phantom equipped with passive and active radiation detectors was exposed outside and at several locations inside the ISS between 2004 and 2011 and measured dose rates between ~0.17 mGy/d and 0.25 mGy/d.

2.3.2 Interplanetary Space

Any spacecraft leaving the Earth's magnetosphere is continuously exposed to the full intensity of GCR and sporadically to SEPs. While astronauts on-board the ISS or any other spacecraft in LEO are effectively protected by Earth itself and its magnetosphere, the only protective mechanism in interplanetary space is the mass shielding provided by the spacecraft itself.

The Radiation Assessment Detector (RAD) (Hassler et al. 2012) of the Mars Science Laboratory (MSL) (Grotzinger et al. 2012) was the first instrument to measure dose rates on a trajectory to Mars between December 2011 and July 2012 during a period of moderate solar modulation. The average dose rate from GCR measured on MSL's cruise was 0.481 ± 0.080 mGy/d in H_2O (0.332 ± 0.023 mGy/d in Si) and 1.84 ± 0.33 mSv/d, corresponding to a quality factor of 3.82 ± 0.25 (Zeitlin et al. 2013). Figure 2.15 illustrates the dose rate in silicon measured by RAD on its interplanetary trajectory to Mars with the underlying relatively constant rates of the GCR and five occurrences of SEP events which manifest in spikes in the dose rate with peak values of up to a few thousand µGy/d.

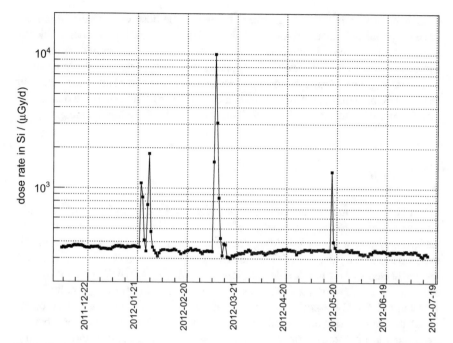

Fig. 2.15 Dose rate measured by the RAD instrument during the transfer of MSL from Earth to Mars (data from Zeitlin et al. 2013)

Semkova et al. (2018) have measured between April and September 2016 at a comparable GCR intensity during the EXO-Mars transit to Mars. The measured GCR dose rates in Si for two different detector configurations were 372 ± 37 µGy/d and 390 ± 39 µGy/d. Other instruments have measured the radiation in lunar transit (Dachev et al. 2011).

These values are approximately factors of 3 to 3.5 greater than what is measured in the COLUMBUS module of ISS during GCR intensity maximum (see Sect. 2.3.1).

These factors contain the influence of the solar modulation, the lack of geomagnetic shielding and shielding by the planet, and differences in the mass shielding provided by the spacecraft and the space station. The most relevant factors are the absence of geomagnetic shielding and the shielding by the planet.

Mass shielding provided by the spacecraft is less effective against GCR and has a major influence mostly on the dose equivalent. The fragmentation of the primary heavy ions leads to a rapidly decreasing quality factor resulting in a drop in dose equivalent for only moderately changing absorbed doses.

A large effort has been undertaken by numerous authors to estimate the radiation exposure from GCR in interplanetary space and the impact and effectiveness of various types of materials (Townsend et al. 1989, 1991; Kim et al. 2010; Mrigakshi et al. 2013b; Slaba et al. 2017; Norbury et al. 2019). Results vary significantly from author to author but are also developing over time. Apart from the shielding geometry and the

GCR boundary condition, the geometry for which the dose was calculated matters. Recently published values by Norbury et al. (2019) for solar minimum under unshielded conditions are 2.5 mSv/d for the dose equivalent in a slab and 1.2 mSv/d for the effective dose equivalent which impressively shows the effect of the self-shielding of the body reducing the exposure by about a factor of two. These results are compatible with results by Kim et al. (2010) and Mrigakshi et al. (2013b). The calculated quality factors are approximately 3 but reducing significantly with increasing shielding.

2.3.3 Moon

Moon lacks both major mechanisms that protect life on Earth from cosmic radiation: an atmosphere and a magnetic field. Astronauts in lunar orbit or on the lunar surface are, except for the shielding provided by any natural or man-made shelter, exposed to the full intensity of GCR and SEPs. In their report on Life Science studies performed during the Apollo missions (NASA 1973), NASA derived average dose values for the different missions between 0.16 rad (1.6 mGy, Apollo 7) and 1.14 rad (11.4 mGy, Apollo 14) and concluded that "radiation was not an operational problem during the Apollo Program." These values include contributions from radiation belt crossings and transfer to a lunar orbit. Most of the dose, however, is contributed by GCR in the lunar orbit or on the surface of the Moon. The report, however, also stated that "it is possible that flares, with the accompanying energetic nuclear particles, might hinder future flights beyond the magnetosphere of the Earth." The lack of an atmosphere means that astronauts which are outside of any habitat, shelter, or vehicle are exposed to the full spectrum of energetic particles and only protected by their space suit.

Due to the absence of atmospheric and magnetic shielding, the exposure to cosmic radiation on the lunar surface is expected to be approximately 50% of the exposure in interplanetary space under comparable shielding conditions if albedo radiation produced in the regolith is neglected. At a given point on the surface, the exposure could be further reduced by nearby rocks, cliffs, crater rims, or other geological formations that reduce the solid angle of open sky. Current measurements are limited to lunar orbit or mission integrated doses from the Apollo missions which also include the transit to the Moon. The first dose rate measurements will be published soon by the Lunar Neutron and Dosimetry (LND) experiment on China's Chang'E 4 lander which started measuring on the lunar surface in January 2019.

Close to the GCR intensity maximum in 2009, the CRaTER instrument on the Lunar Reconnaissance Orbiter Schwadron et al. (2012) measured a dose rate in silicon of 16.5 cGy/year (452 µGy/d) at 10000 km away from the Moon which would result in approximately 8.5 cGy/year (226 µGy/d) on the surface. About one year earlier, the RADOM instrument on Chandrayaan-1 (Dachev et al. 2011) had measured 12.76 µGy/h (306 µGy/d) in silicon (Si) during lunar transfer and 9.46 µGy/h (227 µGy/d) in a 100 km orbit in early 2009. About 34% of the sky are blocked by the Moon at that altitude which results in an approximate dose rate of 171 μGy/d on the surface. These values are projections of measurements in orbit using simple geometrical considerations. The potential contributions of albedo particles are not considered. Numerical simulations of the albedo radiation estimate a 10% to 25% contribution to the dose or effective dose equivalent (Slaba et al. 2011; Reitz et al. 2012; Spence et al. 2013).

Based on the GCR intensity maximum in late 2009 Reitz et al. (2012) also estimated the organ absorbed dose rates (dose equivalent rates) for an astronaut in a space suit on the lunar surface to reach values between 0.16 mGy (0.44 mSv/d) and 0.22 mGy/d (0.82 mSv/d); the corresponding mean quality factors are between $Q \approx 2.4$ and $Q \approx 4.3$. The corresponding effective dose equivalent was estimated to be 0.6 mSv/d.

2.3.4   Mars

The radiation field at the Martian surface is of great interest as Mars is a potential destination for human missions in the near future, as well. In addition to the transit times in interplanetary space, scenarios for a human mission to Mars typically include a stay on the surface of several months. During this time, the astronauts will be continuously exposed to GCR and its secondary radiation field produced in interactions with the Martian atmosphere and regolith and sporadic SEPs. The atmosphere of Mars consists mostly of CO_2, with some contribution of nitrogen (N), argon (Ar) and trace gases and provides shielding against cosmic radiation corresponding to an areal density of 18–23 g/cm^2, depending on altitude, season, and time of day. This shielding is a mass equivalent of approximately 20 cm of water and significantly lower than the protection that Earth's atmosphere provides, which is about a factor of 50 greater at sea level and still a factor of 10 or more greater at commercial flight altitudes. Nevertheless, the atmospheric shielding on Mars is sufficient to alter the primary GCR field drastically. A large fraction of the heavy ions of the GCR suffers fragmentation before it reaches the surface which leads to a significant decrease in the contribution of heavy ions to the dose and the intensity of high-LET particles and a simultaneous drop in the quality factor. On the other hand, a secondary radiation field develops which contains a substantial amount of secondary neutrons with a high quality factor.

The Radiation Assessment Detector (RAD) on the Mars Science Laboratory mission measured a decrease in the quality factor from $Q = 3.82 \pm 0.30$ in cruise to 3.05 ± 0.26 on the surface. The corresponding measured absorbed dose and dose equivalent rates on the Martian surface were 0.21 ± 0.04 mGy/d and 0.64 ± 0.12 mSv/d.

Recently, there has been a substantial effort to compare and improve numerical models for the prediction of the radiation field and exposure on the Martian surface using RAD data (Matthiä et al. 2016; de Wet and Townsend 2017; Flores-McLaughlin

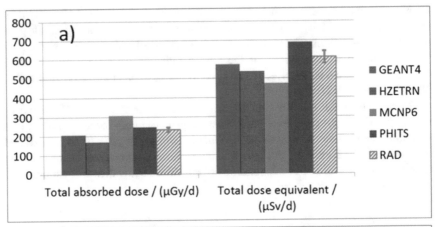

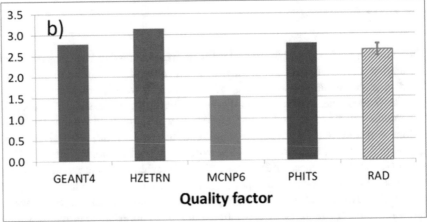

Fig. 2.16 Dose rates (**a**) and corresponding quality factor (**b**) from GCR as measured by the RAD instrument and calculated with different particle transport codes on the surface of Mars for the period between 15 Nov 2015 and 15 Jan 2016. (From Matthiä et al. 2017)

2017; Matthiä and Berger 2017; Ratliff et al. 2017; Slaba and Stoffle 2017). Figure 2.16 summarizes the absorbed dose and dose equivalent rates and the corresponding quality factor of several models in comparison to the RAD measurements. Contributions of different particle types to the absorbed dose and dose equivalent as predicted by GEANT4 model calculations are illustrated in Fig. 2.17. The importance, especiallEq. 2.7y for the dose equivalent, of neutrons produced in the atmosphere and the Martian regolith is evident.

The dose rates in Fig. 2.16 are calculated for a slab of tissue and the values for dose rates in human organs in the identical radiation field are expected to be lower due to the self-shielding effect of the body.

Simonsen et al. (1990) estimated a skin dose equivalent of approximately 0.31 mSv/d to 0.36 mSv/d on the Martian surface. Applying fluence to dose

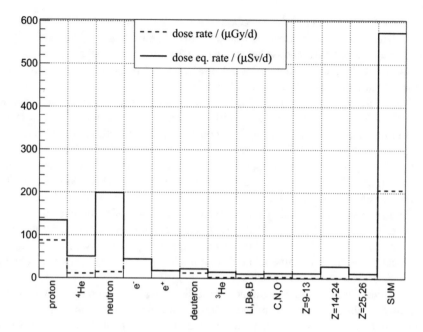

Fig. 2.17 Calculated absorbed dose rates and dose equivalent rates from GCR on the surface of Mars in a slab of tissue from different types of particles (From Matthiä and Berger (2017)) The results have been calculated with GEANT4 for the period between 15 Nov 2015 and 15 Jan 2016

equivalent conversion factors (ICRP 2013) to the results by Matthiä and Berger (2017) gives values of 0.43 mSv/d for the skin and 0.37 mSv/d for the sex-averaged effective dose equivalent.

2.3.5   Solar Particle Events

Solar particle events (SPEs) are highly variable in intensity, energy range, and duration and they are unpredictable. The exposure from each individual event is again strongly dependent on a number of factors, e.g., location of the spacecraft, mass shielding by spacecraft, atmosphere or habitat, magnetic shielding in a planetary magnetosphere, etc. Under such conditions, it is obvious that general conclusions are difficult to draw. Possible ways to address the issue is using a probabilistic approach or defined reference events. Townsend et al. (2018) proposed such a reference event as design basis for missions beyond LEO based on the sum of the October 1989 events. The authors estimated the dose for such an event under unshielded conditions to reach more than one Gy-Eq for both male and female blood-forming organs (BFO, Sect. 3.4) which is a factor of four above the current NASA limit for short-term exposure. They estimated that aluminum shielding of 12 g/cm^2 (for a male) or 15 g/cm^2 (for a female) or polyethylene shielding of 8 g/cm^2 (for a male) or 10 g/cm^2 (for a female) were necessary to reduce the exposure to not exceed the limit. Kim et al. (2017) concluded in their investigation that 5 g/cm^2 shielding is sufficient to keep the BFO dose below the limit for most events but also, that for rare events a shielding above 15 g/cm^2 is necessary. For EVA conditions, they estimated a total event BFO dose of more than 800 mGy-Eq.

During the recent GLE in Sep 2017, Schwadron et al. (2018) estimated a dose of 0.8–0.9 Gy on the lunar surface, while the same event resulted in a dose of 0.418 mGy in silicon measured by RAD on Mars (Zeitlin et al. 2018). An analysis of the event shows that this significantly lower dose measured on the Martian surface is mostly due to the atmospheric shielding but also caused by different event characteristics, i.e., later onset and different spectral shape (Matthiä et al. 2018). Measurements on the ISS showed even lower values caused by a combined mass and magnetospheric shielding effect (68 µGy and 146 µGy for two differently shielded positions, c.f. (Berger et al. 2018). The event is an excellent example showing how the same event can lead to a wide range of different doses, depending on the local shielding environment but also on the location of the observer relative to the source of the energetic particles.

Generally, in LEO and under atmospheric shielding, even such light shielding as provided by the Martian atmosphere, SPEs are of little relevance for human exposure. Only very few strong events have been observed on MIR and ISS in LEO (Berger et al. 2018). Even the exceptionally strong events in September/October 1989 resulted in a combined measured dose of approximately 35 mGy only. This value is expected to decrease significantly for deeper lying organs; it follows that the BFO dose for the event was at least a factor 10 lower than the short-term exposure limit.

References

Adriani O, Barbarino GC, Bazilevskaya GA, Bellotti R, Boezio M, Bogomolov EA, Bonechi L, Bongi M, Bonvicini V, Borisov S, Bottai S, Bruno A, Cafagna F, Campana D, Carbone R, Carlson P, Casolino M, Castellini G, Consiglio L, De Pascale MP, De Santis C, De Simone N, Di Felice V, Galper AM, Gillard W, Grishantseva L, Jerse G, Karelin AV, Koldashov SV, Krutkov SY, Kvashnin AN, Leonov A, Malakhov V, Malvezzi V, Marcelli L, Mayorov AG, Menn W, Mikhailov VV, Mocchiutti E, Monaco A, Mori N, Nikonov N, Osteria G, Palma F, Papini P, Pearce M, Picozza P, Pizzolotto C, Ricci M, Ricciarini SB, Rossetto L, Sarkar R, Simon M, Sparvoli R, Spillantini P, Stozhkov YI, Vacchi A, Vannuccini E, Vasilyev G, Voronov SA, Yurkin YT, Wu J, Zampa G, Zampa N, Zverev VG (2011) PAMELA measurements of cosmic-ray proton and helium spectra. Science 332(6025):69–72
Aguilar M, Aisa D, Alpat B, Alvino A, Ambrosi G, Andeen K, Arruda L, Attig N, Azzarello P, Bachlechner A (2015a) Precision measurement of the helium flux in primary cosmic rays of rigidities 1.9 GV to 3 TV with the alpha magnetic spectrometer on the international space station. Phys Rev Lett 115(21):211101
Aguilar M, Aisa D, Alpat B, Alvino A, Ambrosi G, Andeen K, Arruda L, Attig N, Azzarello P, Bachlechner A (2015b) Precision measurement of the proton flux in primary cosmic rays from rigidity 1 GV to 1.8 TV with the alpha magnetic spectrometer on the international Space Station. Phys Rev Lett 114(17):171103

Badhwar GD, Atwell W, Reitz G, Beaujean R, Heinrich W (2002) Radiation measurements on the Mir Orbital Station. Radiat Meas 35(5):393–422

Baker DN, Panasyuk MI (2017) Discovering Earth's radiation belts. Phys Today 70(12):46–51

Baker D, Erickson P, Fennell J, Foster J, Jaynes A, Verronen P (2018) Space weather effects in the Earth's radiation belts. Space Sci Rev 214(1):17

Band D, Matteson J, Ford L, Schaefer B, Palmer D, Teegarden B, Cline T, Briggs M, Paciesas W, Pendleton G (1993) BATSE observations of gamma-ray burst spectra. I-spectral diversity. Astrophys J 413:281–292

Benton EV (1984) Summary of current radiation dosimetry results on manned spacecraft. Adv Space Res 4(10):153–160

Benton EV, Henke RP (1983) Radiation exposures during space flight and their measurement. Adv Space Res 3(8):171–185

Berger T (2008) Radiation dosimetry onboard the international Space Station ISS. Z Med Phys 18(4):265–275

Berger T, Hajek M (2008) TL-efficiency—overview and experimental results over the years. Radiat Meas 43(2):146–156

Berger T, Hajek M, Summerer L, Vana N, Akatov Y, Shurshakov V, Arkhangelsky V (2004) Austrian dose measurements onboard Space Station MIR and the international Space Station— overview and comparison. Adv Space Res 34(6):1414–1419

Berger T, Bilski P, Hajek M, Puchalska M, Reitz G (2013) The MATROSHKA experiment: results and comparison from extravehicular activity (MTR-1) and intravehicular activity (MTR-2A/2B) exposure. Radiat Res 180(6):622–637

Berger T, Burmeister S, Matthiä D, Przybyla B, Reitz G, Bilski P, Hajek M, Sihver L, Szabo J, Ambrozova I (2017) DOSIS & DOSIS 3D: radiation measurements with the DOSTEL instruments onboard the Columbus laboratory of the ISS in the years 2009–2016. J Space Weather Space Clim 7:A8

Berger T, Marsalek K, Aeckerlein J, Hauslage J, Matthiä D, Przybyla B, Rohde M, Wirtz M (2019) The German Aerospace Center M-42 radiation detector—a new development for applications in mixed radiation fields. Rev Sci Instrum 90(12):125115

Blasi P (2013) The origin of galactic cosmic rays. Astron Astrophys Rev 21(1):70

Boezio M, Carlson P, Francke T, Weber N, Suffert M, Hof M, Menn W, Simon M, Stephens SA, Bellotti R, Cafagna F, Castellano M, Circella M, De Marzo C, Finetti N, Papini P, Piccardi S, Spillantini P, Ricci M, Casolino M, De Pascale MP, Morselli A, Picozza P, Sparvoli R, Barbiellini G, Bravar U, Schiavon P, Vacchi A, Zampa N, Mitchell JW, Ormes JF, Streitmatter RE, Golden RL, Stochaj SJ (1999) The cosmic-ray proton and helium spectra between 0.4 and 200 GV. Astrophys J 518(1):457–472

Boezio M, Carlson P, Francke T, Weber N, Suffert M, Hof M, Menn W, Simon M, Stephens S, Bellotti R (2000) The cosmic-ray electron and positron spectra measured at 1 AU during solar minimum activity. Astrophys J 532(1):653

Buecker H, Horneck G, Allkofer OC, Bartholoma KP, Beaujean R, Cuer P, Enge W, Facius R, Francois H, Graul EH, Henig G, Heinrich W, Kaiser R, Kuhn H, Massue JP, Planel H, Portal G, Reinholz E, Ruther W, Scheuermann W, Schmitt R, Schopper E, Schott JU, Soleilhavoup JP, Wollenhaupt H (1973) The biostack experiment on Apollo 16. Life Sci Space Res 11:295–305

Caffrey JA, Hamby DM (2011) A review of instruments and methods for dosimetry in space. Adv Space Res 47(4):563–574

Cane HV, Lario D (2006) An introduction to CMEs and energetic particles. Space Sci Rev 123(1):45–56

Carlson P (2012) A century of cosmic rays. Phys Today 65(2):30–36

Dachev T (2018) Relativistic electron precipitation bands in the outside radiation environment of the international space station. J Atmos Solar-Terr Phys 177:247–256

Dachev T, Tomov B, Matviichuk YN, Dimitrov P, Vadawale S, Goswami J, De Angelis G, Girish V (2011) An overview of RADOM results for earth and moon radiation environment on Chandrayaan-1 satellite. Adv Space Res 48(5):779–791 de Wet W, Townsend L (2017) A calculation of the radiation environment on the martian surface, Life Sci Space Res 14:51-56.

Desai M, Giacalone J (2016) Large gradual solar energetic particle events. Living Rev Sol Phys 13(1):3

English RA, Benson RE, Bailey JV, Barnes, CM (1973) Apollo experience report: protection against radiation. NASA TN D-7080

Flores-McLaughlin J (2017) Spherical volume radiation transport simulation of the martian GCR surface flux and dose estimation with PHITS, Life Sci Space Res 14:36-42.

Ginet G, O'Brien T, Huston S, Johnston W, Guild T, Friedel R, Lindstrom C, Roth C, Whelan P, Quinn R (2013) AE9, AP9 and SPM: new models for specifying the trapped energetic particle and space plasma environment. In: The van Allen probes mission. Springer, Boston, MA, pp 579–615

Gopalswamy N, Mäkelä P, Yashiro S, Xie H, Akiyama S, Thakur N (2015) High-energy solar particle events in cycle 24. J Phys Conf Ser 642:012012

Grotzinger JP, Crisp J, Vasavada AR, Anderson RC, Baker CJ, Barry R, Blake DF, Conrad P, Edgett KS, Ferdowski B, Gellert R, Gilbert JB, Golombek M, Gomez-Elvira J, Hassler DM, Jandura L, Litvak M, Mahaffy P, Maki J, Meyer M, Malin MC, Mitrofanov I, Simmonds JJ, Vaniman D, Welch RV, Wiens RC (2012) Mars science laboratory mission and science investigation. Space Sci Rev 170(1–4):5–56

Hassler DM, Zeitlin C, Wimmer-Schweingruber RF, Bottcher S, Martin C, Andrews J, Bohm E, Brinza DE, Bullock MA, Burmeister S, Ehresmann B, Epperly M, Grinspoon D, Kohler J, Kortmann O, Neal K, Peterson J, Posner A, Rafkin S, Seimetz L, Smith KD, Tyler Y, Weigle G, Reitz G, Cucinotta FA (2012) The radiation assessment detector (RAD) investigation. Space Sci Rev 170(1–4):503–558

Hess WF (1912) Über Beobachtungen der durchdringenden Strahlung bei sieben Freiballonfahrten. Phys Z XIII:1084–1091

Hess WN (1964) The effects of high altitude explosions. NASA TN D-2402

ICRP (1991) ICRP Publication 60. The 1990 Recommendations of the International Commission on Radiological Protection. Ann. ICRP 21 (1-3). Smith H (Ed.), Pergamon Press Oxford, New York, Frankfurt, Seoul, Sydney, Tokio

ICRP (2007) ICRP Publication 60. The 2007 Recommendations of the International Commission on Radiological Protection. ICRP Publication 103. Ann. ICRP 37(2-4):1-332. Valentin J (Ed.), Elsevier, Orlando, Amsterdam, Tokyo, Singapore

ICRP (2009) ICRP publication 110: adult reference computational phantoms. Joint ICRP/IRCU report. Ann ICRP 39(2):1–165. Clement CH (Ed.), Elsevier, St. Louis, Oxford, Tokyo, Singapore

ICRP (2013) ICRP publication 123. Assessment of radiation exposure of astronauts in space. Ann ICRP 42(4):1–339. Clement CH, Sasaki M (Eds.), Elsevier

ICRP (2016) ICRP publication 132. Radiological protection from cosmic radiation in aviation. Ann ICRP 45(1):1–48. Clement CH, Hamada N (Eds.), SAGE Publications, London, Thousand Oaks, CA, New Delhi, Singapore, Washington DC and Melbourne

ICRU (1986) Report 40: the quality factor in radiation protection. J ICRU os21(1):1–2

ICRU (2011) Report 85: fundamental quantities and units for ionizing radiation. J ICRU 11(1):1–31

Jentsch V (1981) On the role of external and internal source in generating energy and pitch angle distributions of inner-zone protons. J Geophys Res Space Phys 86(A2):701–710

Jiggens P, Heynderickx D, Sandberg I, Truscott P, Raukunen O, Vainio R (2018) Updated model of the solar energetic proton environment in space. J Space Weather Space Clim 8:A31

Kim M-HY, Hayat MJ, Feiveson AH, Cucinotta FA (2009) Prediction of frequency and exposure level of solar particle events. Health Phys 97(1):68–81

Kim M-HY, Qualls GD, Slaba TC, Cucinotta FA (2010) Comparison of organ dose and dose equivalent for human phantoms of CAM vs. MAX. Adv Space Res 45(7):850–857

Kim M-HY, Blattnig SR, Clowdsley MC, Norman RB (2017) Using spectral shape and predictor fluence to evaluate temporal dependence of exposures from solar particle events. Space Weather 15(2):374–391

Kireeva SA, Benghin VV, Kolomensky AV, Petrov VM (2007) Phantom—dosimeter for estimating effective dose onboard international space station. Acta Astronaut 60(4):547–553

Konradi A, Atwell W, Badhwar GD, Cash BL, Hardy KA (1992) Low earth orbit radiation dose distribution in a phantom head. Int J Rad Appl Instrum D 20(1):49–54

Lishnevskii A, Panasyuk M, Benghin V, Petrov V, Volkov A, Nechaev OY (2012) Variations of radiation environment on the international Space Station in 2005–2009. Cosm Res 50(4):319–323

Ludwig GH (1962) The NASA program for particles and fields research in space. NASA TN D-1173

Matthiä D, Berger T (2017) The radiation environment on the surface of Mars—numerical calculations of the galactic component with GEANT4/PLANETOCOSMICS. Life Sci Space Res 14:57–63

Matthiä D, Ehresmann B, Lohf H, Köhler J, Zeitlin C, Appel J, Sato T, Slaba T, Martin C, Berger T, Boehm E, Boettcher S, Brinza DE, Burmeister S, Guo J, Hassler DM, Posner A, Rafkin SCR, Reitz G, Wilson JW, Wimmer-Schweingruber RF (2016) The Martian surface radiation environment—a comparison of models and MSL/RAD measurements. J Space Weather Space Clim 6:A13

Matthiä D, Hassler DM, de Wet W, Ehresmann B, Firan A, Flores-McLaughlin J, Guo J, Heilbronn LH, Lee K, Ratliff H, Rios RR, Slaba TC, Smith M, Stoffle NN, Townsend LW, Berger T, Reitz G, Wimmer-Schweingruber RF, Zeitlin C (2017) The radiation environment on the surface of Mars—summary of model calculations and comparison to RAD data. Life Sci Space Res 14:18–28

Matthiä D, Meier MM, Berger T (2018) The solar particle event on 10–13 September 2017: spectral reconstruction and calculation of the radiation exposure in aviation and space. Space Weather 16(8):977–986. https://doi.org/10.1029/2018SW001921

McKenna-Lawlor S, Bhardwaj A, Ferrari F, Kuznetsov N, Lal AK, Li Y, Nagamatsu A, Nymmik R, Panasyuk M, Petrov V, Reitz G, Pinsky L, Muszaphar Shukor S, Singhvi AK, Straube U, Tomi L, Townsend L (2014) Feasibility study of astronaut standardized career dose limits in LEO and the outlook for BLEO. Acta Astronaut 104(2):565–573

Mrigakshi AI, Matthiä D, Berger T, Reitz G, Wimmer-Schweingruber RF (2013a) How galactic cosmic ray models affect the estimation of radiation exposure in space. Adv Space Res 51(5):825–834

Mrigakshi AI, Matthiä D, Berger T, Reitz G, Wimmer-Schweingruber RF (2013b) Estimation of galactic comic ray exposure inside and outside the Earth's magnetosphere during the recent solar minimum between solar cycles 23 and 24. Adv Space Res 52(5):979–989

Narici L, Berger T, Matthiä D, Reitz G (2015) Radiation measurements performed with active detectors relevant for human space exploration. Front Oncol 5:273

NASA (1973) Biomedical results of APOLLO

NASA (1991) Radiation protection for human missions to the Moon and Mars

NASA (1999) Solar cycle variations and application to the space radiation environment

NASA (2013) Space radiation cancer risk projections and uncertainties—2012

NASA (2014) NASA space flight human-system standard volume 1, Revision A: Crew Health

NASA (2015) Badhwar—O'Neill 2014 galactic cosmic ray flux model description

NCRP (2000) NCRP report 132: radiation protection guidance for activities in low-earth orbit

Norbury JW, Slaba TC, Aghara S, Badavi FF, Blattnig SR, Clowdsley MS, Heilbronn LH, Lee K, Maung KM, Mertens CJ, Miller J, Norman RB, Sandridge CA, Singleterry R, Sobolevsky N, Spangler JL, Townsend LW, Werneth CM, Whitman K, Wilson JW, Xu SX, Zeitlin C (2019) Advances in space radiation physics and transport at NASA. Life Sci Space Res 22:98–124

Powell C (1950) The cosmic radiation. Nobel Lecture

Puchalska M, Bilski P, Berger T, Hajek M, Horwacik T, Korner C, Olko P, Shurshakov V, Reitz G
(2014) NUNDO: a numerical model of a human torso phantom and its application to effective
dose equivalent calculations for astronauts at the ISS. Radiat Environ Biophys 53(4):719–727

Ratliff HN, Smith MBR, Heilbronn LH (2017) Simulation of the GCR spectrum in the Mars
Curiosity rover's RAD detector using MCNP6, Life Sci Space Res 14:43-50.

Reames DV (1999) Particle acceleration at the sun and in the heliosphere. Space Sci Rev
90(3):413–491

Reames DV (2013) The two sources of solar energetic particles. Space Sci Rev 175(1):53–92

Reitz G, Berger T, Bilski P, Facius R, Hajek M, Petrov V, Puchalska M, Zhou D, Bossler J, Akatov
Y, Shurshakov V, Olko P, Ptaszkiewicz M, Bergmann R, Fugger M, Vana N, Beaujean R,
Burmeister S, Bartlett D, Hager L, Palfalvi J, Szabo J, O'Sullivan D, Kitamura H, Uchihori Y,
Yasuda N, Nagamatsu A, Tawara H, Benton E, Gaza R, McKeever S, Sawakuchi G, Yukihara E,
Cucinotta F, Semones E, Zapp N, Miller J, Dettmann J (2009) Astronaut's organ doses inferred
from measurements in a human phantom outside the international space station. Radiat Res
171(2):225–235

Reitz G, Berger T, Matthiae D (2012) Radiation exposure in the moon environment. Planet Space
Sci 74(1):78–83

Richmond RG (1972) Radiation dosimetry for the Gemini program. NASA TN D-6695

Sato T, Endo A, Sihver L, Niita K (2011) Dose estimation for astronauts using dose conversion
coefficients calculated with the PHITS code and the ICRP/ICRU adult reference computational
phantoms. Radiat Environ Biophys 50(1):115–123

Schaefer HJ, Sullivan JJ (1976) Atlas of nuclear emulsion micrographs from personnel dosimeters
of manned space missions. NASA CR-149446

Schaefer HJ, Benton EV, Henke RP, Sullivan JJ (1972) Nuclear track recordings of the astronauts'
radiation exposure on the first lunar landing mission Apollo XI. Radiat Res 49(2):245–271

Schwadron NA, Baker T, Blake B, Case AW, Cooper JF, Golightly M, Jordan A, Joyce C, Kasper
J, Kozarev K, Mislinski J, Mazur J, Posner A, Rother O, Smith S, Spence HE, Townsend LW,
Wilson J, Zeitlin C (2012) Lunar radiation environment and space weathering from the cosmic
ray telescope for the effects of radiation (CRaTER). J Geophys Res Planets 117(E12):E00H13

Schwadron NA, Rahmanifard F, Wilson J, Jordan AP, Spence HE, Joyce CJ, Blake JB, Case AW,
Wet W, Farrell WM, Kasper JC, Looper MD, Lugaz N, Mays L, Mazur JE, Niehof J, Petro
N, Smith CW, Townsend LW, Winslow R, Zeitlin C (2018) Update on the worsening particle
radiation environment observed by CRaTER and implications for future human deep-space
exploration. Space Weather 16(3):289–303

Semkova J, Koleva R, Benghin V, Dachev T, Matviichuk Y, Tomov B, Krastev K, Maltchev S,
Dimitrov P, Mitrofanov I, Malahov A, Golovin D, Mokrousov M, Sanin A, Litvak M, Kozyrev
A, Tretyakov V, Nikiforov S, Vostrukhin A, Fedosov F, Grebennikova N, Zelenyi L, Shurshakov
V, Drobishev S (2018) Charged particles radiation measurements with Liulin-MO dosimeter
of FREND instrument aboard ExoMars trace gas Orbiter during the transit and in high elliptic
Mars orbit. Icarus 303:53–66

Simonsen LC, Nealy JE, Townsend LW, Wilson JW (1990) Space radiation-dose estimates on the
surface of mars. J Spacecraft and Rockets 27(4):353–354

Simpson JA (1983) Elemental and isotopic composition of the galactic cosmic rays. Ann Rev Nuc
Part Sci 33:323–382

Simpson JA (2000) The cosmic ray nucleonic component: the invention and scientific uses of the
neutron monitor. In: Cosmic rays and earth. Springer, Dordrecht, pp 11–32

Slaba TC, Blattnig SR (2014) GCR environmental models I: sensitivity analysis for GCR environments. Space Weather 12(4):217–224

Slaba TC, Stoffle NN (2017) Evaluation of HZETRN on the Martian surface: sensitivity tests and
model results. Life Sci Space Res 14:29–35

Slaba TC, Blattnig SR, Clowdsley MS (2011) Variation in lunar neutron dose estimates. Radiat
Res 176(6):827–841

Slaba TC, Bahadori AA, Reddell BD, Singleterry RC, Clowdsley MS, Blattnig SR (2017) Optimal shielding thickness for galactic cosmic ray environments. Life Sci Space Res 12:1–15

Spence HE, Golightly MJ, Joyce CJ, Looper MD, Schwadron NA, Smith SS, Townsend LW, Wilson J, Zeitlin C (2013) Relative contributions of galactic cosmic rays and lunar proton "albedo" to dose and dose rates near the moon. Space Weather 11(11):643–650

Straube U, Berger T, Reitz G, Facius R, Fuglesang C, Reiter T, Damann V, Tognini M (2010) Operational radiation protection for astronauts and cosmonauts and correlated activities of ESA medical operations. Acta Astronaut 66(7):963–973

Townsend LW, Nealy JE, Wilson JW, Atwell W (1989) Large solar flare radiation shielding requirements for manned interplanetary missions. J Spacecraft Rockets 26(2):126–128

Townsend LW, Shinn JL, Wilson JW (1991) Interplanetary crew exposure estimates for the august 1972 and October 1989 solar particle events. Radiat Res 126(1):108–110

Townsend LW, Adams JH, Blattnig SR, Clowdsley MS, Fry DJ, Jun I, McLeod CD, Minow JI, Moore DF, Norbury JW, Norman RB, Reames DV, Schwadron NA, Semones EJ, Singleterry RC, Slaba TC, Werneth CM, Xapsos MA (2018) Solar particle event storm shelter requirements for missions beyond low earth orbit. Life Sci Space Res 17:32–39

Tylka AJ, Dietrich WF, Atwell W (2010) Assessing the space-radiation hazard in ground-level enhanced (GLE) solar particle events. 2010 Fall AGU Meeting, San Francisco, CA

Uchihori Y, Fujitaka K, Yasuda N, Benton E (2002) Intercomparison of radiation instruments for cosmic-ray with heavy ion beams at NIRS (ICCHIBAN project). J Radiat Res 43(Suppl):S81–S85

UNSCEAR (2000) Sources and effects of ionizing radiation, ANNEX B, Exposures from natural radiation sources. UNSCEAR 2000 REPORT, New York 1: 97–99

Van Allen JA, Frank LA (1959) Radiation around the earth to a radial distance of 107,400 km. Nature 183(4659):430–434

Van Allen JA, McIlwain CE, Ludwig GH (1959a) Radiation observations with satellite 1958 ε. J Geophys Res 64(3):271–286

Van Allen JA, McIlwain CE, Ludwig GH (1959b) Satellite observations of electrons artificially injected into the geomagnetic field. J Geophys Res 64(8):877–891

Vernov SN, Chudakov AE (1960) Investigations of cosmic radiation and of the tab—errestrial corpuscular radiation by means of rockets and satellites. Sov Phys Usp 3(2):230–250

Vette JI (1991) The NASA/national space science data center trapped radiation belt model (1964–1991). National Space Science Data Center (NSSDC), Greenbelt, MD

Warren CS, Gill WL (1964) Radiation dosimetry aboard the spacecraft of the eight Mercury-Atlas mission (MA-8). NASA TN D-1862

Wilson JW (1978) Environmental geophysics and SPS shielding

Xapsos MA, Summers GP, Barth JL, Stassinopoulos EG, Burke EA (2000) Probability model for cumulative solar proton event fluences. IEEE Trans Nuc Sci 47(3):486–490

Yasuda H (2009) Effective dose measured with a life size human phantom in a low earth orbit mission. J Radiat Res 50(2):89–96

Yasuda N, Uchihori Y, Benton ER, Kitamura H, Fujitaka K (2006) The intercomparison of cosmic rays with heavy ion beams at NIRS (ICCHIBAN) project. Radiat Prot Dosim 120(1–4):414–420

Zeitlin C, Hassler DM, Cucinotta FA, Ehresmann B, Wimmer-Schweingruber RF, Brinza DE, Kang S, Weigle G, Böttcher S, Böhm E, Burmeister S, Guo J, Köhler J, Martin C, Posner A, Rafkin S, Reitz G (2013) Measurements of energetic particle radiation in transit to Mars on the Mars science laboratory. Science 340:1080–1084

Zeitlin C, Hassler DM, Guo J, Ehresmann B, Wimmer-Schweingruber RF, Rafkin S, Freiherr von Forstner JL, Lohf H, Berger T, Matthiä D, Reitz G (2018) Analysis of the radiation hazard observed by RAD on the surface of mars during the September 2017 solar particle event. Geophys Res Lett 45:5845–5851

Chapter 3
Radiation in Space: The Biology

Abstract The galactic cosmic radiation (GCR) results constant exposure of astronauts to charged particles of various energies at a low-dose rate. A traversal of a charged particle through a cell nucleus can result in (complex) DNA damage which initiates the DNA damage response (DDR). During this response, the cell might arrest in the cell cycle in order to gain time for DNA repair. Depending on damage severity, cell type, and other factors, different outcomes such as cell death, premature differentiation, senescence, or chromosomal aberrations are possible. In addition to effects on the main target molecule, the DNA, non-targeted effects contribute to overall outcome for a GCR exposed organism. Compared to X- or γ-rays, heavy charged particles can have a high relative biological effectiveness (RBE) for inducing different biological outcomes, including cancer-relevant outcomes and cancer in rodents. Primarily, late tissue sequels like genetic alterations, cancer and non-cancer effects, i.e., cataracts and degenerative diseases, e.g., of the central nervous system, are potential risks for space travelers. Cataracts were observed to occur with earlier onset and more frequently in astronauts exposed to even low doses of GCR. In addition, the continuously existing risk of acute exposure to high proton fluxes during a solar particle events (SPE) implies to threaten immediate survival of the astronauts in case of insufficient shielding by eliciting the acute radiation syndrome (ARS).

Keywords DNA damage · DNA repair · Cell death · Cell cycle · Senescence · Chromosomal aberrations · Relative biological effectiveness · Acute radiation syndrome · Cancer · Cataract · CNS effects

3.1   Introduction

During space missions, astronauts experience a chronic whole body exposure with in average low dose of single energetic particles (electrons, protons, α-particles, heavy ions, and neutrons) at low dose rate. This chronic exposure results locally and temporally in an inhomogeneous dose distribution in the body tissues. While some cells hit by an energetic heavy ion contribute with a high dose to the total perceived dose, not-hit cells hardly receive any dose at all (Reitz and Hellweg 2018).

Additionally to this chronic exposure from galactic cosmic radiation (GCR), an acute whole body exposure can occur as a consequence of Solar Particle Events (SPE) delivering high radiation doses at a high dose rate during a relatively short time period (Reitz and Hellweg 2018). For interplanetary missions, this imposes a risk for mission success and astronaut health, while in low Earth orbit (LEO), where the ISS is operating, this exposure is low to moderate (Sect. 2.3 in Chap. 2), because the geomagentic shielding is fairly efficient. An overview of space radiation exposure is given in Fig. 3.1.

The limited knowledge of the biological effects from exposure to heavy charged particles is an ongoing concern in human spaceflight (Cucinotta and Chappell 2011). The exposure level during human missions in LEO and beyond LEO in the interplanetary space and to other planets is different in total dose and radiation quality (Sect. 2.3 in Chap. 2). The extent, type, and onset of radiation effects observed in mammals depend on the dose, the dose rate, the radiation quality, and the individual sensitivity of the exposed human or animal.

Acute radiation effects in humans or other mammals appear quite soon after exposure to a high dose in a short period of time (minutes to a few days). Late effects, such as cancer, can occur after years or decades in survivors of radiation exposure and a threshold dose is not known. Late effects have an occurrence probability proportional to the level of exposure (Reitz and Hellweg 2018).

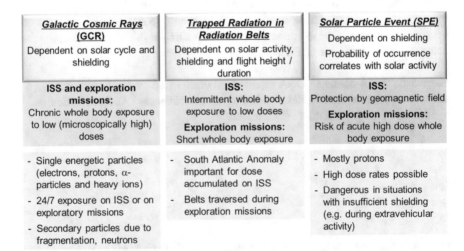

Fig. 3.1 Radiation exposure of astronauts during space missions in LEO and beyond

The basic molecular mechanisms how radiation-induced damage and cellular stress impacts cell fate decisions which might pave the way for carcinogenesis (Sridharan et al. 2016) and other health effects are described in the next subchapter.

3.2   Molecular Mechanisms of Heavy Charged Particles Effects

The molecular events start with the deposition of the radiation energy in the cell. The distribution pattern of this energy deposition is concentrated in a track. Molecules in the track such as deoxyribonucleic acid (DNA) are damaged. The cell reacts actively by initiating a DNA damage response resulting in DNA repair, alterations in gene expression, changes in cell cycle progression, or even cell death (Fig. 3.2). Cellular changes resulting from damage to the main target of radiation, DNA, in a directly hit cell are called "targeted effects."

3.2.1   Energy Deposition by Heavy Ions in Biological Matter

When traveling through matter, heavy ions continuously deposit energy. This results in ionizations along the track of the particle. The track consists of a core with a high density of ionizations and penumbra with a lower ionization density. This

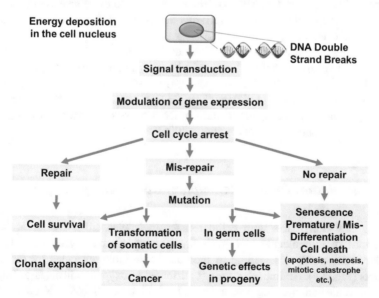

Fig. 3.2 Targeted effects and downstream cellular responses after exposure to ionizing radiation, adapted from Prise (2006) and Reitz and Hellweg (2018)

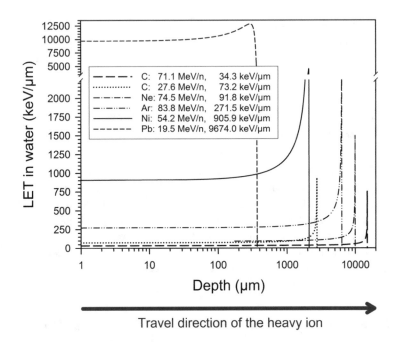

Fig. 3.3 Bragg curves visualizing the linear energy transfer (LET) of different heavy ions (Pb—lead, Ni—nickel, Ar—argon, Ne—neon, C—carbon) with energies from 19.5 to 83.8 MeV/n at the entrance point with increasing depth in water (Hellweg 2012)

micrometer broad lateral extension of the track is due to low-linear energy transfer (LET) energetic electrons (δ-rays) (Krämer and Kraft 1994, Cucinotta et al. 2000, Durante and Cucinotta 2008).

Depending on the initial energy of the particle, the energy deposition remains constant for a specific traveling distance ("plateau") and increases sharply before all residual energy is deposited ("Bragg peak") (Fig. 3.3). This energy deposition pattern is specific for charged particle radiation and responsible for local concentration of damages and a distinct spatial pattern of DNA lesions (Brenner and Ward 1992, Rydberg 1996, Ponomarev and Cucinotta 2006).

3.2.2 Radiation-Induced (Complex) DNA Damage

Ionizing radiation reacts with cellular macromolecules either by direct (direct ionizations of biomolecules) or by indirect action via radiolysis (ionization of cellular water molecules leading to modifications of biomolecules by radiolysis products (Fig. 3.4). An important role is attributed to oxygen radicals attacking the DNA molecule and inducing base damage, loss of bases, or DNA–DNA and DNA–protein crosslinks. Disruption of the DNA ribose-phosphate backbone leads to

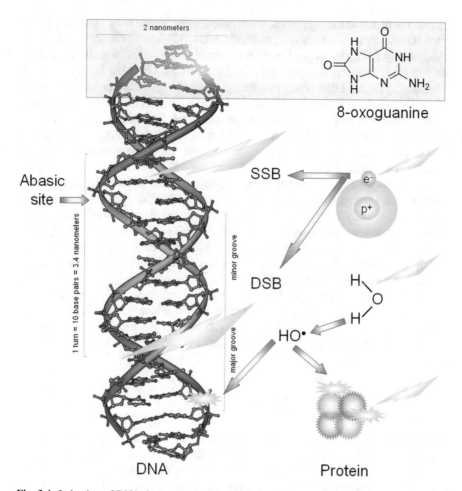

Fig. 3.4 Induction of DNA damage by ionizing radiation. Direct energy deposition in the DNA molecule or indirect damage after radiolysis of water molecules to hydroxyl and other radicals result in the formation of single-strand breaks (SSB), double-strand breaks (DSB), base damages such as oxidized guanine (8-oxoguanine), loss of bases with creation of an abasic site or DNA–protein crosslinks. DNA and 8-oxoguanine molecule were available at Wikimedia Commons (Michael Störk and Ed/commons.wikimedia.org/)

single- and double-strand breaks (SSB, DSB), dependent on the proximity of induced ionizations (Reitz and Hellweg 2018). Indirect action via water-derived radicals (reactive oxygen species, ROS) plays an important role in the induction of biological effects by low-LET radiation, including X- and γ-rays. These ROS are very short-lived and have a small diffusion range. Direct action makes a larger contribution to the biological effectiveness of high-LET radiation than indirect action does (Hirayama et al. 2009).

Low- and high-LET radiation induces a different spatial distribution of direct DNA damage (Sridharan et al. 2016) and of ROS (Goodhead 1988), resulting in clusters of different damage (SSB, DSB, base lesions, abasic sites, etc.) called complex DNA damage (formerly: multiply damaged sites) (Eccles et al. 2010, Asaithamby and Chen 2011). Complex damage is uncommon for endogenous ROS or low-LET radiation (Durante and Cucinotta 2008). Complex DNA damage is challenging for the DNA repair machinery, as short fragments can arise and multiple repair pathways have to be coordinated (Cucinotta and Durante 2006, Sridharan et al. 2016). The delay in repair of such complex DNA damage may result in persisting damage until DNA replication. The fidelity of complex lesion repair is expected to be lower compared to simple lesions (Ward 1994, Stenerlow et al. 2000, Schöllnberger et al. 2004).

Any DNA damage can change the information in the affected DNA molecule, with different consequences. DNA damage can lead to mutations and cell transformations. Especially DSB are considered as the central element in cell inactivation (killing) by ionizing radiation (Jackson and Bartek 2009) (Reitz and Hellweg 2018) and as key precursors of most early and late radiation effects (Durante and Cucinotta 2008). Unrepaired complex DNA damage can target cells for growth arrest, cell death, or cellular senescence (Sridharan et al. 2016).

Misrejoining of DNA ends from radiation-induced DSBs can result in chromosomal aberrations such as deletions or translocations (Cucinotta and Durante 2006, Sridharan et al. 2016). Mutations can initiate the multistep carcinogenic process (Cucinotta and Durante 2006, Durante and Cucinotta 2008).

3.2.3 DNA Damage Response

Radiation effects on the organismal level are based on cellular reactions. The cellular response to radiation is predominantly a DNA Damage Response (DDR) that detects lesions, signals their presence, and promotes their repair (Jackson and Bartek 2009). This signal transduction pathway involves multiple sensors for different types of DNA lesions, transducer molecules, and a variety of effector molecules and enzymes for repair (Fig. 3.5). The DDR results in potentially cell-protective (cell cycle arrest, DNA repair, survival), cell-altering (senescence, mutations), or even cell-destructive responses (different types of cell death) (Fig. 3.7) (Khanna et al. 2001). Signal transduction leads to the activation of multiple pathways and various transcription factors (transducers), resulting in the expression of certain genes whose protein products (effectors) are involved in these responses (Hellweg et al. 2016).

A central role in coordinating the radiation response plays the phosphatidylinositol kinase-related protein ATM, product of the ATM gene that is mutated in patients with *Ataxia telangiectasia* (AT). Fibroblasts from AT patients were 3–4 fold more sensitive to ionizing radiation than those from normal donors (Higurashi and Conen 1973). The serine/threonine kinase ATM is considered as DNA damage sensor that binds directly or with the help of other proteins to damaged DNA or to replication protein A (RPA)-coated single-stranded DNA (Khanna et al. 2001, Jackson and Bartek 2009).

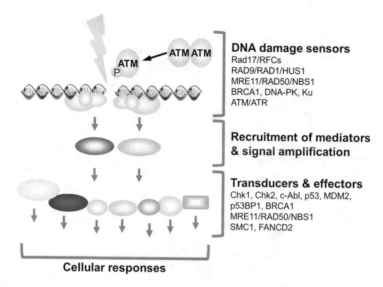

Fig. 3.5 Model of the DNA damage response (DDR). Various sensor proteins recognize the presence of a lesion in the DNA, e.g., a DNA double-strand break (DSB) induced by ionizing radiation (yellow arrow). These sensors trigger signaling pathways that initiate a wide variety of cellular responses (Fig. 3.7) via mediators, transducers, and finally effectors. Adapted from Jackson and Bartek (2009) and Wilson (2004)

3.2.3.1 Repair of Radiation-Induced DNA Damage

To safeguard their function as information carriers, damaged DNA molecules have to be repaired while all other macromolecules in the cell can be replaced if they are damaged. In dividing cells, repair has to be completed before damages are fixed as mutations during DNA replication and cell division. DNA damage that is not repaired with high fidelity can lead to mutations and chromosomal aberrations or to mitotic cell death. Therefore, cells need to slow down or stop progression through the cell cycle after DNA damage occurred. In mammalian cells, the activation of DNA repair and arrests in different phases of the cell cycle are immediate responses to DNA damage (Wang and Cho 2004), and these responses are designed to protect a damaged cell from damage fixation and to promote its recovery.

Within several minutes after damage induction, a multitude of DNA repair proteins is recruited to the damaged site and forms the so-called Ionizing Radiation-Induced Foci (IRIF) (Bekker-Jensen and Mailand 2010). In response to DNA DSB formation, the MRN (MRE11-RAD50-NBS1) complex recruits the DNA damage sensor ATM. ATM is activated by autophosphorylation at Serine 1981 (Shiloh 2003) and phosphorylates then many other proteins directly or indirectly. Substrates for ATM are p53, MDM2, BRCA1, and Rad51 that are involved in homologous recombination (HR), the transcriptional repressor C-terminal binding protein interacting protein (CtIP), NBS1, RAD9, and the serine/threonine protein kinase CHK2 that are involved in cell cycle checkpoint execution (Pawlik and Keyomarsi 2004).

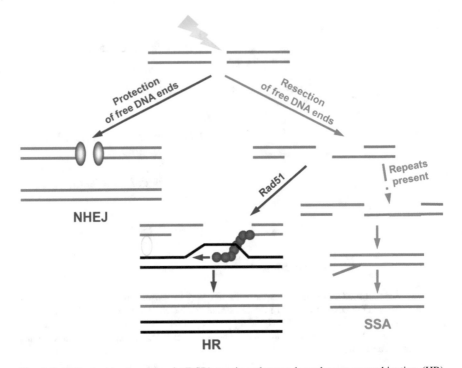

Fig. 3.6 DNA double-strand break (DSB) repair pathways: homologous recombination (HR), non-homologous end-joining (NHEJ), and single-strand annealing (SSA) (adapted from Kass and Jasin 2010). NHEJ joins adjacent free DNA ends with no or little homology (microhomology), irrespectively of their sequence and must therefore be considered to be error-prone. In this pathway, DNA ends are protected from end resection by binding to the Ku heterodimer. HR is initiated by 5' to 3' end resection. Onto the formed 3' single-stranded tail, the RAD51 nucleoprotein filament assembles and invades into a homologous DNA duplex (black), thereby initiating repair synthesis. The newly synthesized strand is then displaced and anneals to the other DNA end (not shown) thereby completing the HR reaction. When such end resection occurs at sequence repeats of the sister chromatid (green lines), an alternative pathway, SSA, can take place. In this case, the complementary single strands anneal at the repeat. This gives rise to a copy number variant. As shown by mutational analysis, factors involved in HR and NHEJ likewise "compete" at steps indicated by the numbers: (1) loss of the canonical NHEJ factors (Ku, DNA ligase IV/XRCC4) leads to increased end resection and hence HR and SSA. (2) End resection mutants (e.g., Sae2) demonstrate increased NHEJ. (3) Disruption of RAD51 filament formation leads DNA ends to be thread into SSA (Kass and Jasin 2010)

Activated ATM rapidly phosphorylates the histone variant H2AX on serine 139 to generate γ-H2AX (Rogakou et al. 1998, 1999). γ-H2AX can be detected as microscopically visible foci with high sensitivity by immunofluorescence and therefore represents a useful marker for the cellular response to DNA DSB (Banath and Olive 2003, Rothkamm and Löbrich 2003). After their assembly, the IRIF initiate repair at the damaged site.

Non-homologous end-joining (NHEJ) and homologous recombination (HR) are the major pathways for the repair of DSBs (Fig. 3.6) (O'Driscoll and Jeggo 2006,

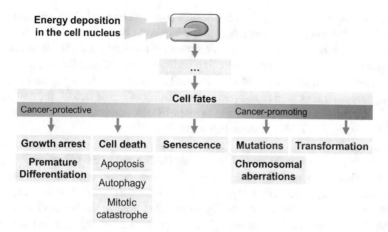

Fig. 3.7 Possible cell fates after radiation exposure. The three dots indicate the DDR shown in Fig. 3.5

Wyman and Kanaar 2006, Löbrich and Jeggo 2007, Phillips and McKinnon 2007). Compared to X-rays, the repair foci induced by heavy ions are larger and more dynamic in the first 2 h after irradiation, and they persist longer (Costes et al. 2006, Asaithamby et al. 2008). With near horizontal orientation of the cells towards the heavy ion beam, streaks of foci indicate the path of a particle through the cell nucleus (Desai et al. 2005). The slower DNA repair kinetics and the incomplete repair of DNA DSB induced by high-LET radiation (Baumstark-Khan et al. 2003, Asaithamby et al. 2008) are generally explained by the higher complexity of the DNA lesions (Goodhead et al. 1993).

NHEJ is the predominant DSB repair pathway in mammalian cells and "glues" free DNA ends together. As it is independent from template information, it is error-prone and chromosomal translocations, deletions, or insertions can arise. When key proteins of this pathway, such as Ku, DNA-PKcs, XRCC4, or ligase IV are mutated, mammalian cells become exceptionally sensitive to ionizing radiation and defective in V(D)J recombination[1] of the immunoglobulin and the T-cell receptor (TCR) genes (DeFazio et al. 2002, van der Burg et al. 2009).

In mammalian cells, HR depends on the presence of an intact sister chromatid as template, and therefore DSB repair occurs only when duplicated chromatin is present in the S- and G2 phase of the cell cycle. It retrieves the information from the undamaged sister chromatid and can therefore be error-free. HR is initiated by resection of DNA ends at the DSB sites. The accrued long stretches of single-stranded DNA (ssDNA) invade duplex DNA with a homologous sequence and repair synthesis starts (Mimitou and Symington 2009).

---

[1] V(D)J recombination of the immunoglobulin and T-cell receptor (TCR) genes is the prerequisite for differentiation of mature B- and T-lymphocytes from their lymphoid precursors. During V(D)J recombination, gene segments encoding Variable (V), Diversity (D) and Joining (J) segments of immunoglobulin (VDJ), and TCR (VJ) genes are rearranged to close adjacency for transcription and translation.

This newly synthesized strand dissociates from the invaded DNA and anneals to the other DNA end. If sequence homology exists on both sides of the DNA DSB, single strand annealing (SSA) of the complementary single strands can occur, with removal of the resulting flaps by the flap endonuclease (Rad10/ERCC1 and Rad1) and therefore, loss of genetic information will be the result (Kass and Jasin 2010).

Numerous factors such as temporary availability of repair factors, the cell's position in the mitotic cell cycle, and the extent of DNA end resection affect the cell's decision to repair a DSB via these pathways.

The repair of complex DNA damage might require the involvement of other repair pathways such as base excision repair (BER) and nucleotide excision repair (NER) (Shikazono and O'Neill 2009). BER can repair base damage such as 8-oxoguanine. In BER, a DNA glycosylase recognizes a damaged base and removes it before nuclease, polymerase, and ligase proteins finalize the repair (Jackson and Bartek 2009). Larger helix-distorting lesions are repaired by the NER pathway with faster kinetics for the transcribed strand of active genes (transcription-coupled repair) than for inactive genome regions (global genome repair) (Tornaletti 2009).

Several space experiments were performed addressing the question of a possible effect of microgravity on DNA repair processes. In the stick insect *Carausius morosus*, a synergistic effect of cosmic radiation and microgravity was observed (Bücker et al. 1986a, b). The eggs in different developmental stages were exposed to spaceflight conditions (including microgravity), and controls were kept on a 1 × g centrifuge. The cosmic radiation hits in eggs were detected by particle track detector foils that were placed between the egg layers (Biostack). After hatching, more abnormalities occurred after heavy ion hits to the eggs under microgravity conditions than under 1 × g centrifugal force (Bücker et al. 1986a).

In order to determine whether such microgravity effects are also relevant in human cells, the space experiment KINETICS was performed during the SpaceLab Mission International Microgravity Laboratory 2 (IML-2) in the Biorack of ESA. Human fibroblasts were irradiated on ground, kept in a non-metabolic state, and incubated in orbit at 37 °C to initiate repair processes. No significant differences in the DNA DSB repair kinetics on ground and in orbit were observed. However, it is not clear, whether all repair processes were inhibited in the non-metabolic state or whether some of the DNA DSB were repaired under normal gravity conditions before transfer of irradiated cells to orbit (Horneck et al. 1996, 1997). Since then, a few more space experiments and several ground-based experiments addressed the microgravity and DNA repair question (Manti 2006, Lu et al. 2017) and the results remain conflicting. A growth-stimulating effect of microgravity and changes in gene expression might be a strong contributor to microgravity effects on the DDR (Moreno-Villanueva et al. 2017). Further space experiments using an ionizing radiation source on the ISS will help to solve this question. The use of an inflight centrifuge to achieve 1 × g will allow to segregate microgravity effects from the effects of other ISS environmental factors, such as the elevated carbon dioxide levels (Law et al. 2014) which can be associated with oxidative DNA damage (Lu et al. 2007).

3.2.3.2 Reactive Oxygen Species (ROS)

The ROS produced directly by ionizing radiation disappear quickly, but in the following, ROS can be released from the mitochondria of irradiated cells (Leach et al. 2001). They act as signaling intermediaries, and this signaling information is integrated with the nuclear signals converging, e.g., in activation of Nuclear Factor κB (NF-κB) (Wu and Miyamoto 2007, Ghosh and Hayden 2008, Deorukhkar and Krishnan 2010, Sylvester et al. 2018). ROS may remain elevated for days and weeks up to years (Durante and Cucinotta 2008), e.g., in immortalized human bronchial epithelial cells exposed to iron ions (Werner et al. 2014), human neural stem cells exposed to protons and iron ions (Tseng et al. 2013, 2014, Baulch et al. 2015), in iron ion irradiated hippocampal precursor cells (Limoli et al. 2007), in the bone marrow of proton-irradiated mice (Chang et al. 2015), and in intestinal epithelial cells of whole body iron ion irradiated mice (Datta et al. 2012). Persisting DNA damage and damaged mitochondria are suggested to drive the secondary ROS generation after the initial insult (Kawamura et al. 2018, Murray et al. 2018).

3.2.3.3 Activation of Transcription Factors and Gene Expression

The activation of several signal transduction pathways by ionizing radiation results in altered expression of series of target genes. The promoters or enhancers of these genes may contain binding sites for one or more transcription factors, and a specific transcription factor can influence the transcription of multiple genes.

A central element of the mammalian radiation response is activation of the transcription factor p53 (TP53). p53 has a short half-life and is stabilized in response to a variety of cellular stresses including ionizing radiation after phosphorylation by ATM (Fei and El-Deiry 2003). The target genes include CDKN1A which protein product p21 provokes a cell cycle arrest and ribonucleotide reductase RRM2B (p53R2) that increases cellular levels of deoxyribonucleotides for efficient DNA repair (Karagiannis and El-Osta 2004). p53 can also induce apoptosis via Bax and related proapoptotic proteins (Karagiannis and El-Osta 2004). According to Gudkov and Komarova (2010a, b), the response initiated by p53 is tissue-specific, producing severe damage in tissues prone to p53-dependent apoptosis, such as the hematopoietic system (thereby contributing to the acute radiation syndrome, Sect. 3.4 in Chap. 3), hair follicles, and oligodendroblasts in the spinal cord after whole body irradiation of mice. Cell cycle arrest and activation of DNA repair occur in the vascular endothelial cells of the small intestine, while connective tissues and epithelial cells react with a growth arrest to whole body irradiation.

The transcription factor NF-κB is strongly activated by heavy ions (Hellweg et al. 2011) and upregulates the expression of several cyto- and chemokines (Chishti et al. 2018). It can therefore induce intercellular communication and inflammatory responses (Hellweg 2015), which might favor tumor promotion and progression (Habelhah 2010). Antiapoptotic genes also belong to the target gene spectrum of NF-κB, but currently, there are no indications that these genes are upregulated after heavy ion exposure and NF-κB did not confer a survival advantage for heavy ion irradiated cells (Chishti et al. 2018, Hellweg et al. 2018).

The increase of ROS after radiation exposure could elicit an antioxidant response via the nuclear erythroid-derived 2-related factor 2 (Nrf2) pathway (McDonald et al. 2010), and an endoplasmic response to stress (UPR response) may complement the complex cellular response (Gueguen et al. 2019). The involvement transcription factors in the cellular response to heavy ions was recently reviewed (Hellweg et al. 2016).

The changes in gene expression induced by ionizing radiation via the abovementioned transcription factors depend on dose, dose rate, time after irradiation, radiation quality, cell type, cell cycle phase, and possibly on other factors. CDKN1A, GADD45α, and NER pathway genes (XPC) were consistently upregulated by ionizing radiation as an immediate response, setting the course for cell cycle arrest and repair, normal function, senescence, or apoptosis (Snyder and Morgan 2004).

3.2.4   Cell Fates After Radiation Exposure

Integration of nuclear, cytoplasmic and membrane signals, and the outcome of the repair process lead to different potential cell fates, including growth arrest, loss of stemness and premature differentiation, cell death via different mechanisms, senescence, mutations, and/or transformation (Fig. 3.7).

3.2.4.1   Radiation-Induced Cell Cycle Arrests

Cell cycle checkpoints are important surveillance systems to maintain genomic integrity. The term checkpoint was coined to describe a mechanism that monitors the order of events in the cell cycle to ensure that a cell cycle event occurs only after a prior event has been completed properly (Hartwell and Weinert 1989). Checkpoints delay the cell cycle if the DNA is damaged to avoid replication and segregation of damaged DNA. Cells initiate DNA damage checkpoints after genotoxic stress exposure to procure time for repair in dividing cells (Pawlik and Keyomarsi 2004, Krempler et al. 2007). Effectors of the cell cycle checkpoints downstream of ATM are cyclin-dependent kinases (CDK) complexed with a regulatory Cyclin (Iliakis et al. 2003). The G1/S checkpoint prevents replication of damaged DNA. Ionizing radiation can initiate an arrest at this checkpoint only in cells with wild-type p53 function (Hwang and Muschel 1998). In response to DNA damage during the S-phase of the cell cycle, replicative DNA synthesis can be stopped or slowed down (S phase arrest) (Khanna et al. 2001). The G2/M checkpoint blocks the entry into mitosis if cells incur damage in G2 or earlier phases of the cell cycle (Pawlik and Keyomarsi 2004). It thereby prevents the segregation of damaged chromosomes during mitosis. The release from G2/M checkpoint following exposure to ionizing radiation depends upon the level of residual DSB—a permanent arrest in G2 is only sustained when the damage level remains above a certain threshold of 10–20 DSB (Di Leonardo et al. 1994, Krempler et al. 2007). The longer persistence of DNA damage after high-LET radiation exposure might explain the observed prolonged cell cycle arrests induced by high-LET (Fournier and Taucher-Scholz 2004) in comparison to low-LET irradiation. The arrest in the G1 phase of the cell cycle (Scholz et al. 1994) and the G2 arrest depend on LET (Blakely et al. 1989).

3.2.4.2   Radiation-Induced Cell Death and Death Bypass Mechanisms

In case of complete and error-free repair of radiation-inflicted DNA damage, survival of the cell is expected. If the attempts of the cell to repair or bypass the radiation-induced DNA damage fail, cell death can be a consequence. Misrepair of the DNA damage can result in mutations, chromosomal aberrations, and delayed cell death. Exposure of human cells to ionizing radiation can provoke different types of cell death (e.g., apoptosis, necrosis, autophagy, or mitotic catastrophe). Low- and high-LET radiation show different potencies in induction of cell death.

Apoptotic cell death has been shown to be a central factor in radiation-induced death of cells from radiosensitive tissues, such as cells of the myeloid and lymphoid lineage, acinar cells from the parotid gland, thymocytes, and spermatogonia (Radford et al. 1994, Hendry et al. 1995). Fibroblasts are more prone to terminal differentiation after radiation exposure (Rodemann et al. 1991), and radiation-induced apoptosis occurs only in small percentage of cells (<10%), with higher rates for high- compared to low-LET radiation (Aoki et al. 2000).

High doses (20 Gy) of ionizing radiation are required to induce classical necrosis with cell swelling, dilatation of organelles, random DNA degradation, and leakage through the damaged membrane in mammalian cells (Cornelissen et al. 2002, Pawlik and Keyomarsi 2004).

Mitotic catastrophe is assumed to be the major cell death pathway in solid tumors following radiotherapy and is accompanied by hyperamplification of the centrioles in the centrosome or microtubule organizing center, which are responsible for spindle formation (Eriksson and Stigbrand 2010). This results in multipolar mitosis followed by the formation of micronuclei containing mis-segregated chromosomes (Muller et al. 1996) or of multiple nuclei contained in a giant cell (Eriksson and Stigbrand 2010).

Autophagy is a cellular catabolic process that involves sequestration of cytoplasmic portions into vesicles (autophagosomes) that then fuse with lysosomes to allow degradation of their contents for the purpose of recycling cellular components to sustain metabolism during nutrient deprivation and to prevent accumulation of damaged proteins and organelles (Gudkov and Komarova 2010a, b). Relatively high doses (6–30 Gy) are necessary to initiate detectable autophagy (Hellweg 2012).

3.2.4.3 Radiation-Induced Senescence

Ionizing radiation can trigger cellular senescence (Qin et al. 2018). Senescent cells are functionally impaired and no longer able to divide but remain metabolically active for long periods of time (Nakanishi et al. 2009). Cellular senescence is regarded as anti-cancer mechanism, and evidence for its involvement in aging and age-related diseases is accumulating (Rodier and Campisi 2011). It is also considered as important contributor to radiotherapy-induced tumor regression (Eriksson and Stigbrand 2010). Senescent cells express senescence-associated β-galactosidase, precursor protein of Alzheimer's β-amyloid peptide, degradative enzymes, cytokines (e.g., IL-6 and IL-8), and growth factors which might be tumor-promoting (Li et al. 2018). This excretion profile was called senescence-associated secretory phenotype (Coppe et al. 2010) and is caused at least partly by increased NF-κB activity (Freund et al. 2011).

3.2.4.4 Mutations and Chromosomal Aberrations

The induction of mutations by radiation of different LET was examined in several mammalian cell systems, especially at the hypoxanthine phosphoribosyltransferase (HPRT) locus of Chinese Hamster Ovary (CHO) cells. Mutations are late endpoints of DNA damage and are only detectable in the survivors of the irradiation. Chromosomal aberrations are microscopically visible DNA alterations arising from the repair of DSBs (Obe et al. 2002). Following ionizing radiation exposure, un- or misrepaired DSBs, and chromosome mis-seggregation can cause aberrations in chromosome structure and number (Cornforth 2006, Durante and Cucinotta 2008, Sridharan et al. 2016), such as the formation of dicentrics (chromosomes with two centromeres) and acentric fragments that are usually observed as micronuclei. Heavy ions are extremely effective in eliciting chromosomal exchanges, up 30 times more effective compared to low-LET radiation during interphase (George et al. 2003, Durante and Cucinotta 2008). Complex aberrations lead to cell death (Durante and Cucinotta 2008).

The chromosomal aberration test allows analysis of the aberrations in human peripheral lymphocytes that are collected by venipuncture. A significant increase of chromosomal aberrations was observed in the lymphocytes of eight astronauts who participated in long-duration NASA/Mir missions (Letaw et al. 1989, George et al. 2001). The half-life of translocations in blood lymphocytes of different astronauts displayed a high inter-individual variability (George et al. 2005). In a study involving MIR and ISS astronauts, the effects in ISS crews were small and no significant effects were observed in visiting astronauts (Obe et al. 1997). For a Mars mission, an increase of dicentrics frequencies in peripheral lymphocytes of 10–40 times above background level was predicted (Obe et al. 1999).

As densely ionizing radiation qualities such as heavy ions are thought to produce a large fraction of (complex) chromosome interchanges (Johannes et al. 2004, Horstmann et al. 2005a, b), such lesions might be observed in blood lymphocytes of astronauts after space missions. Using high-resolution multicolor banding *in situ* hybridization (mBAND) of chromosome 5 to assess the frequency of intrachromosomal exchanges, no inversions were detected in blood lymphocytes of 11 astronauts after short- or long-term stays in LEO (MIR and ISS missions) (Horstmann et al. 2005a, b). Analysis via multicolor fluorescence *in situ* hybridization (mFISH) revealed also no complex type exchanges (Horstmann et al. 2005a, b).

3.2.5 Non-Targeted Effects

The responses of cells to radiation in the absence of direct DNA damage are designated as non-targeted effects (Prise et al. 2006). They usually follow a nonlinear dose–effect relationship and include bystander effects in which charged particle hit cells transmit signals to surrounding cells by direct contact (via gap junctions) or by secreting factors (Marin et al. 2015). Resulting changes the tissue microenvironment encompass oxidative damage (Mothersill and Seymour 2004, Rola et al. 2005, Sokolov et al. 2007), persistent inflammation (Kusunoki and Hayashi 2008), and extracellular matrix remodeling (Park et al. 2003, Durante and Cucinotta 2008). When irradiation of one tissue or organ affects responses in another unirradiated tissue or organ, these are called abscopal effects (Cucinotta et al. 2019).

Genomic instability is characterized by an increased rate of acquisition of alterations in the mammalian genome (chromosomal aberrations, aneuploidy, delayed lethal mutations, gene amplifications, micronuclei) in the progeny of irradiated cells (Hamada et al. 2008, Marin et al. 2015, Tang and Loke 2015). This phenomenon was observed, e.g., after high-LET α-particle exposure (Kadhim et al. 1992, Sabatier et al. 1992) and was explained by centrosome deregulation (Maxwell et al. 2008). While its role in radiation carcinogenesis was under debate (Huang et al. 2003, Sieber et al. 2003, Kadhim et al. 2006), recent results show the involvement of non-targeted effects in heavy ion-induced mammary carcinogenesis (Barcellos-Hoff and Mao 2016).

Adaptive responses represent another type of non-targeted effects. They were observed in various model systems after fractionated radiation exposure: a low priming dose (0.01–0.5 Gy) protects from effects of a subsequent higher dose (called challenging dose) which is explained by activation of protective pathways (Joiner et al. 2001, Schöllnberger et al. 2004, Tapio and Jacob 2007). A radiation-induced adaptive response can be maintained for a few hours to several months, as indicated by a reduction of cell death, chromosomal aberrations, mutations, and radiosensitivity or induction of DNA repair (Tapio and Jacob 2007). Low priming doses might result in one track per cell nucleus in average, meaning that part of the cells were untargeted (Tapio and Jacob 2007).

Another effect can occur after low-dose exposure: low-dose hypersensitivity which means that cells die extensively from exposure to a single low dose (<0.1 Gy) which cannot be extrapolated from the survival at higher doses (Tapio and Jacob

2007). It was explained as increased radiation sensitivity when DNA repair is not fully functional (Tapio and Jacob 2007).

The role of non-targeted effects in the *in vivo* radiation response, in the pathogenesis of radiation-induced diseases, in inter-individual radiation sensitivity and in radiation risk is difficult to assess (Morgan 2003, Prise et al. 2006, Kadhim et al. 2013).

3.3   Relative Biological Effectiveness

The biological effects of high-LET radiation show quantitative and in some aspects qualitative differences compared to the low-LET radiation response. Comparison of the biological effects of different radiation qualities is usually being performed in terms of relative biological effectiveness (RBE). The RBE of the radiation quality in question (test radiation) is defined as the ratio of a dose D_x of the reference radiation (for example, X-rays) and a dose D_t of a test radiation (for example, heavy ions) that elicit the same biological effect (Durante and Cucinotta 2008). The RBE depends on many parameters, including LET, particle velocity and charge, radiation dose, dose rate and dose fractionation, biological endpoint, type of the irradiated cells or tissues, and oxygen concentration. The RBE is a central parameter for space radiation risk assessment. The RBE was determined for various biological endpoints in different biological systems, usually for single ion species (examples are shown in Table 3.1). The different particle species and energies are usually summarized based on their LET, or based on their charge and speed. The maximal RBE was reached with an LET from 90 to 200 keV/μm for many endpoints (Thacker et al. 1979, Ainsworth et al. 1983, Kraft et al. 1989, Yatagai 2004). In rodent experiments, heavy ions had a higher carcinogenic effectiveness than low-LET radiation (Fry and Storer 1987, Alpen et al. 1993, Dicello et al. 2004, Ando et al. 2005, Durante and Cucinotta 2008).

For the same LET, ions of higher Z will have a higher velocity and, therefore, wider tracks (NCRP 2006). Above 100 keV/μm, the track structure results in individual dependencies of RBE and LET for different ions (Kraft et al. 1989, Stoll et al. 1995, 1996).

RBE values for late effects were the basis to estimate the quality factor Q(LET). At low LET (<10 keV/μm), Q equals 1. Around 100 keV/μm LET, Q reaches 30 and decreases at very high-LET values (overkill effect) (Durante and Cucinotta 2008). The dose equivalent (H) is the product of the energy dose and Q and its unit is Sievert (Sv): $H = Q \times D$ (Sect. 2.2 in Chap. 2).

For assigning Q, the radiation protection committees International Commission on Radiation Protection (ICRP) and National Council on Radiation Protection and Measurements (NCRP) identify the most relevant biological endpoints, collect the RBE at low doses (RBEmax) for these endpoints, and calculate an average RBE (Cucinotta et al. 2005). The RBE reviews are published in ICRP Report 40 (ICRP 1984), NCRP Report 104 (NCRP 1990), and ICRP Report 92 (ICRP 2003). The

Table 3.1   Relative biological effectiveness of charged particles

Biological Endpoint	Model System	Radiation Quality (LET)	RBE	References
Survival	Human kidney T1 cells	α-particles (25–200 keV/μm)	1.5–7.0	Barendsen et al. (1960, 1963)
Survival (D_0)	Human skin fibroblasts	α-particles (100 keV/ μm)	5.2	Chen et al. (1984)
Apoptosis	Human neuronal progenitor cell line (Ntera2)	1 GeV/n iron ion (150 keV/μm)	3.4	Guida et al. (2005)
DNA DSB induction	V79 cells	Protons (11–31 keV/ μm)	~1	Belli et al. (1996)
DNA DSB induction	CHO-K1 cells	Carbon ions (17–400 keV/μm)	~1	Heilmann et al. (1995)
DNA DSB induction	V79 cells	Protons (31 keV/μm) Helium ions (131 keV/μm)	1.9 1.2	Belli et al. (2001)
γ-H2AX foci induction	82-6 hTERT immortalized fibroblast cells	Iron ions (150 keV/ μm)	1.8	Whalen et al. (2008)
NF-κB activation	HEK/293 cells	Argon ions (~270 keV/μm)	~9	Hellweg et al. (2011)
Mutation induction	V79-753B cells, HGPRT locus	Protons (11–24 keV/ μm)	5.0–7.7	Belli et al. (1991)
	Human embryonic fibroblast-like cells (HE20 cells), HPRT locus	Carbon ions (20–310 keV/μm)	2.2–7.5	Suzuki et al. (2003), Yatagai (2004)
Intrachromosome exchanges	Human lymphocytes	Protons (0.4 keV/μm) α-particles (121 keV/ μm) Iron ions (140 keV/ μm)	0.8–1.3 1.9–8.1 1.9–8.9	Wu et al. (1997)
Chromosomal aberrations	Rat bone marrow cells	Iron ions (~150 keV/ μm)	~3.2	Brooks et al. (2001)
Chromosomal aberrations (premature chromosome condensation)	Human lymphocytes	^{1}H-, ^{3}He-, ^{12}C-, ^{40}Ar-, ^{28}Si-, ^{56}Fe-, ^{197}Au ions (0.4–1393 keV/μm)	Up to ~35	George et al. (2003)
Cataracts	Mice	Argon, neon, and carbon ions (<100 keV/μm)	3–5	Jose and Ainsworth (1983), Yang and Ainsworth (1987)
	Rats	Iron ions (190 keV/ μm) and argon ions (88 keV/μm)	50–200 (at 0.01 Gy)	Brenner et al. (1993)
	Rats	Argon ions (88 keV/ μm)	Up to 40	Merriam Jr. et al. (1984)

(continued)

Table 3.1 (continued)

Biological Endpoint	Model System	Radiation Quality (LET)	RBE	References
Transformation per surviving cell	HeLa × human skin fibroblast cell line CGL-1	Carbon ions (26 and 153 keV/μm)	2.5 and 12	Bettega et al. (2009)
	C3H10T1/2 cells	α-particles (~100 keV/μm)	~20	Miller et al. (1995)
	Syrian hamster embryo cells	Carbon and silicon ions (13–400 keV/μm)	2–7	Han et al. (1998)
	Golden hamster embryo cells	Neon ions (530 keV/μm) Helium ions (36 keV/μm) Helium ions (77 keV/μm)	3.3 2.4 3.3	Suzuki et al. (1989)
Tumor induction	Harderian gland tumors in mice (with pituitary isografts)	Iron ions (190 keV/μm) Carbon ions (~40–90 keV/μm)	27 12	Fry et al. (1985)
		Helium ions (1.6 keV/μm) Iron ions (193 keV/μm) Iron ions (253 keV/μm)	2.3 39.6 20.3	Alpen et al. (1993, 1994)
	Mice, legs locally irradiated	Carbon ions (4–300 keV/μm)	2.2	Ando et al. (2005)
	Rats, mammary carcinoma	Carbon ions (40–90 keV/μm), high doses and low doses	2 & ~10	Imaoka et al. (2007)
	Mice, legs locally irradiated	Carbon ions (15, 45, or 75 keV/μm)—high doses	0.87, 1.29, or 2.06	Ando et al. (2014)
	APC[1638N/+] [a] male and female mice, intestinal tumors	Carbon ions (13 keV/μm) Silicon ions (70 keV/μm) Iron ions (150 keV/μm)	1.6 and 1.4 6.4 and 3.7 3.7 and 3.5	Cucinotta et al. (2017)
	APC[1638N/+] male mice, colorectal tumors	Carbon ions (13 keV/μm) Silicon ions (70 keV/μm) Iron ions (150 keV/μm)	5.8 15.0 11.9	

[a]APC[1638N/+] mice are a mouse model of colorectal cancer. The Adenomatous polyposis coli (APC) gene is mutated in Familial Adenomatous Polyposis. Truncation of the APC gene at codon 1638 results in the formation of a few polyps and later adenomas and adenocarcinomas in the small intestine (Smits et al. 1997)

experiments on the induction of tumors of the Harderian gland by heavy ions (LET > 100 keV/μm) are decisive for Q, they revealed a very high RBE (Table 3.1) (Fry et al. 1983, Alpen et al. 1994). The data for protons and heavy ions for tumor-relevant endpoints are very scarce, and the experimental conditions are very variable (reference radiation, doses, dose rates, LET). This results in large uncertainties for RBE of space-relevant radiation qualities (Cucinotta et al. 2005). The RBE of GCR components and SPE protons for induction of acute and late radiation effects has to be known for quantitative risk estimation for astronauts and space travelers (Chap. 4).

3.4   Acute Radiation Effects

Exposure to an acute single dose can elicit the acute radiation syndrome (ARS), also termed radiation sickness. Symptoms depend on different key issues, such as the total absorbed radiation dose, the type and quality of radiation, the dose distribution in body tissues and organs and the individual radiation sensitivity (Cronkite 1964, Donnelly et al. 2010, Drouet and Herodin 2010, Dorr and Meineke 2011) (Reitz and Hellweg 2018) .

The prodromal stage of the ARS is hallmarked by a rapid onset of nausea, vomiting, and malaise (dose >0.5 Gy for sparsely ionizing radiation) followed by a nearly symptom-free dose-dependent latent phase of weeks to days (Reitz and Hellweg, 2018). Stimulation of IL-1β in the caudal medulla by the *Nervus vagus* is considered as the major route for induction of vomiting after acute radiation exposure (Makale and King 1993). The electrical activity and neurochemical metabolism of the central nervous system is disturbed already at doses of 1–2 Gy (Tofilon and Fike 2000). These effects may be involved in the development of prodromal symptoms already at lower doses (Marquette et al. 2003). The prodromal stage can be life threatening, if vomiting occurs while the astronaut performs an extravehicular activity in a space suit.

The tissue with rapid turnover is affect in the first place. For the individuum, the effects result from depletion of the already differentiated compartiments by cell-death inducing mechanisms such as apoptosis (Purgason et al. 2018), and from absent or failing replacement of depleted tissues or organs as a consequence from blocking the mitotic cell cycle and cell death occuring in the stem cell compartment. Only such cells which vanquish the cell cycle block and resume proliferation are able to substitute radiation-damaged tissue to preserve its normal function (Reitz and Hellweg 2018).

One of the most radiosensitive organs in the body is the bone marrow harboring hematopoietic stem cells. Its functional depression (bone marrow or hematopoietic syndrome) occurs in humans after exposure to doses of 0.7–4 Gy. Another radiation sensitive tissue is the intestine, especially the cells in the gut's crypts, responsible for cellular renewal of short lived functional villi cells. Their inactivation (gastrointestinal tract syndrome) occurs after radiation exposure to 5–12 Gy. With higher doses (> 20 Gy) death of cells from the differentiated cell compartment lead to manifestation of the central nervous system syndrome. The dose that leads to death for 50% of the exposed humans within 30 days was estimated to amount to 3–4 Gy (lethal dose: $LD_{50/30}$). Several studies propose an exacerbated response of the innate immune system to play an important role in pathogenesis of all three ARS sub-syndromes (Van der Meeren et al. 2005, Drouet and Herodin 2010, Jacob et al. 2010). Epithelial and endothelial cells are supposed to produce pro-inflammatory cytokines in the course of ARS (Van der Meeren et al. 2005). The predominant feature of the bone marrow syndrome is immune suppression (Reitz and Hellweg 2018).

In the bone marrow syndrome, progressive lymphopenia develops during the first days after radiation exposure. Exposure to ~2 Gy results in a maximal depression of the lymphocytes in the blood (Cronkite 1964). The lymphocyte deprivation decreases the resistance to infections. An immunosuppression is expected after exposure to an absorbed dose of 0.5 Gy (Dainiak et al. 2003). Damage to the mega-karyoblasts in the bone marrow causes thrombocytopenia with the result of increased bleeding tendency. A possible early granulocytosis is followed by a progressive granulocytopenia (Chao 2007). The longer life span of erythrocytes causes a later develpment of anemia (within 2–6 weeks) than lymphopenia. Within days after accidental sublethal whole body γ-radiation, a reduced mitotic index and cytologi-cal abnormalities (such as mitotic bridges, multipolar mitosis, binucleated cells, micronuclei) were observed in human bone marrow cells (e.g., erythroblasts), which have the potential to persist at a lower frequency for years after the accident (Fliedner et al. 1964). If the patient cannot overcome the critical period of the possibly reversible aplastic state of the bone marrow, death occurs from sepsis usually at 30-60 days post-exposure (Cronkite 1964).

The gastrointestinal tract syndrome arises after only a short latent period for whole body irradiation with doses of 5–12 Gy. It is caused by loss of the intestinal epithelium from massive cell destruction and lack of successful mitotic activity in the intestinal epithelium crypts as well as injury to the fine submucosal vascula-ture. This allows bacteria to invade and to produce local and systemic inflammation and blood poisoning (sepsis) with the consequence of multiple organ failure (Drouet and Herodin 2010). Death occurs between 3 and 10 days after exposure (Reitz and Hellweg 2018). In this complex chain of events, endothelial cells and parenchymal cells are damaged (Gaugler et al. 2005), endothelial cells and leukocytes are acti-vated, pro-inflammatory cytokines such as interleukin-8 (IL-8), IL-6, IL-12 and IL-18, and ROS are produced (Singer et al. 2004, Jacob et al. 2010), and neuropep-tides are released (Gourmelon et al. 2005). Activation of the innate immune system was suggested to be involved in target organ damage and adverse metabolic and hemodynamic responses (Jacob et al. 2010). A recent experiment exposing mice to whole body proton irradiation (250 MeV) revealed that the number of apoptotic cells in small intestine crypts increased and the number of surviving crypts decreased already after a dose of 0.1 Gy (Purgason et al. 2018). As a consequence of gene expression changes, a low-grade inflammation of the small intestine at this low dose is expected (Purgason et al. 2018).

The onset of the central nervous system syndrome occurs several hours to days after exposure to very high acute radiation doses (Reitz and Hellweg 2018).

Symptoms include loss of coordination, confusion, convulsions, eventually coma, and signs of the bone marrow and gastrointestinal syndromes. As the prognosis is fatal, survival is impossible. Above 10–12 Gy, 100% lethality has to be expected. In the brain, overexpression of cytokines such as tumor necrosis factor α (TNF-α), IL-1α, and IL-1β occurs within several hours after whole body irradiation of mice (Jacob et al. 2010).

For space missions, exposure of astronauts to mostly protons from a large SPE could lead to trigger ARS. Animal experiments have shown that a simulated SPE (2 Gy of protons delivered within 36 h) induce anemia, indicated by a decreased number of erythrocytes, and immunosuppression with decrease of lymphocytes, monocytes, and granulocytes numbers in blood and of relative spleen mass (Gridley et al. 2008). In mice exposed to 3 Gy of protons, a strong immunodepression was observed already 12 h after exposure, with the nadir on day 4 (Kajioka et al. 2000). The radiosensitivity of the lymphocyte subpopulations decreased in the order B-cells > CD8$^+$ cytotoxic or regulatory T-cells > CD4$^+$ T-helper cells > natural killer (NK) cells (Kajioka et al. 2000). Possible after effects of the immunosuppression are augmented infection susceptibility, induction of cancer by promotion of initiated cells, and immune system disorders such as autoimmunity or hypersensivity (Reitz and Hellweg 2018).

3.5 Chronic and Late Radiation Effects

Chronic or delayed effects of radiation exposure include cancer induction and non-cancer effects. Non-cancer effects include degenerative diseases of the central nervous and the cardiovascular system, the eye lens (cataract), and possibly other organ systems.

3.5.1 Cancer

The induction of cancer is an essential and life-threatening long-term consequence for atomic bomb survivors and victims of radiation accidents, exposed to intermediate and low dose whole body radiation exposure. In radiation therapy, secondary tumors are stochastically induced by low radiation doses in the tissues surrounding the tumor. For astronauts, the risk of developing cancer as late effect is under discussion. Ionizing radiation can definitively cause initiation of tumors, due to its potency to damage DNA and induce mutations in proto-oncogenes or lead to loss of function in tumor-suppressor genes. The ionizing radiation's role in tumor promotion and progression is less well established (Reitz and Hellweg 2018). The probability of tumor induction in the dose range below 1 Sv was derived from the cancer incidence (solid tumors, leukemia) in atomic bomb survivors who were exposed at high dose rates (Pierce and Preston 2000, Pierce et al. 1996, 2012). With increasing survival times of patients after cancer radiotherapy, there is growing concern for the risk of secondary cancer, especially in children who are inherently more radiosensitive (Baskar 2010). Breast, thyroid, colon, and lung are described as major sites for radiation-induced solid cancers (Preston et al. 2007, Durante and Cucinotta 2008). According to the NCRP, the natural cancer incidence differs between males and females, and radiation adds cancer risks for the breast and the ovaries and a higher risk from for lung cancer in females (NCRP 2000, Cucinotta and Durante 2009). This is considered by lower radiation exposure limits for females than for males (NCRP 2000).

During a long-term space exploration mission, astronauts accumulate high exposures to GCR. Human cancer risk data for GCR, especially for heavy ions, are not available, and radiobiological data are limited (Reitz and Hellweg 2018). Therefore, the risk estimates that were derived from terrestrial exposure situations (mostly atomic bomb survivors) and projected to the conditions of interplanetary travel are afflicted with large uncertainties (Durante and Cucinotta 2008).

Linking the early biological effects of heavy ions to the cancer probability is an extremely difficult task. A causal relationship between the occurence of chromosomal aberrations in peripheral blood lymphocytes and development of cancer allows assessing the cancer risk by suitable analysis of the frequency of aberrations (Durante et al. 2001). After long duration space flights a significant increase in chromosomal aberrations was detected in astronauts (George et al. 2001); however, data on cancer mortality and incidence in astronauts are limited and due to small group size and low mission doses before the ISS era, they miss statistical confidence (Committee on the Longitudinal Study of Astronaut Health 2004).

Using scoring of chromosomal aberrations for risk assessment technique, the relative cancer risk for cosmonauts on long-term missions at the former Russian space station Mir was estimated to be ~1.2–1.3 (Durante et al. 2001). For a Mars mission reaching the planet's surface and lasting 1000 days with a mission dose of 1.07 Sv, the uncertainties in cancer risk projections were estimated to be in the order of 400–600% (Cucinotta et al. 2001b)(Cucinotta et al. 2015).

3.5.2   Degenerative Tissue Effects

Degenerative tissue effects could result from the chronic GCR exposure or from acute exposure(s) to SPEs. Degenerative effects in the eye (cataract) after heavy ion exposure were in the focus of research since the 1980s.

NASA funded many projects on heavy ions effects on the central nervous system during the last 20 years. Also cardiovascular pathologies are considered to be of main concern in NASA's Human Research Program Integrated Research Plan (Milstead et al. 2019).

Multiple factors, including lifestyle-associated factors (e.g., obesity, alcohol, and tobacco) are involved in the pathogenesis of degenerative diseases, making the assessment of space radiation-associated risks particularly difficult. Especially in the lower dose ranges, population-based risk estimates can be obscured by these factors. Large uncertainties in the low-dose range and a strong healthy worker effect in astronauts are expected.

3.5.2.1 Cataract

After exposure to protracted sparsely ionizing radiation the threshold for cataract formation is discussed to be 2 Gy, some investigators even suggest doses as low as 100–300 mGy, or dismiss any threshold (Worgul et al. 1999). Studies on rabbits exposed to neon or argon ions, let Lett et al. (1980) conclude that astronauts possibly experience late radiation effects decades after a long-duration space mission beyond LEO. For mice and rats exposed to densely ionizing radiation (energetic heavy ions), it was shown that such radiation qualities are especially effective in cataract formation (Table 3.1), even at doses below 2 Gy (Worgul et al. 1989, Worgul, Brenner et al. 1993, Hall et al. 2006). ATM heterozygosity is accompanied by higher sensitivity to cataractogenesis (Hall et al. 2006, Durante and Cucinotta 2008). Blakely et al. (2010) published about 100 mGy as lower threshold for cataract induction. Furthermore, contrarily to other radiation effects, lowering the dose rate does not reduce the cataract risk (Hamada 2017). In Rhesus monkeys (*Macaca mulatta*), with a median life span ~24 years, formation of cataracts was observed after the exposure to protons of different energies (~20 years), and also in rabbits after exposure to highly energetic iron ions (Lett et al. 1991, Cox et al. 1992). In addition to animal experiments, investigations of patients undergoing radiotherapy provide valuable data for the risk assessment of space radiation-induced cataract formation or cancer induction in astronauts (Blakely et al. 1994, Lett et al. 1994, Wu et al. 1994). So far, the occurence of lens opacities has been the only proven late effect of space radiation exposure in astronauts (Cucinotta et al. 2001a, Rastegar et al. 2002). Given an overall mission death risk of some percent for human space flights and the fact that cataract surgery can restore vision, it is not justified to consider cataractogenesis as a major critical health risk in short- to medium-term space flight. For long-term missions that span several years, the expression time for cataracts must be taken into account. During such a mission, a cataract can evolve before returning to our home planet Earth and thereby it poses a risk for the success of the mission (Reitz and Hellweg 2018).

The eye lens is highly radiosensitive as it lacks mechanisms to remove damaged cells, and its transparency depends on correct differentiation of lens epithelial cells into lens fibers. Abnormal differentiation, slow or incomplete repair of DNA DSBs, senescence and crystalline changes appear to be responsible for the high radiosensitivity of the lens (Hamada 2017).

3.5.2.2 Central Nervous System Effects

Central nervous system (CNS) effects of GCR are currently under intense discussion as "space-brain" and include cognitive deficits and neurological damage (Jandial et al. 2018) observed after irradiation of mice and rats with heavy ions using various assays (Shukitt-Hale et al. 2003), e.g., Morris water maze (Shukitt-Hale et al. 2000, 2007, Higuchi et al. 2002, Villasana et al. 2013), the Barnes maze (Britten et al. 2012, 2016, Wyrobek and Britten 2016), the 8-arm radial maze (Denisova et al. 2002), Y-maze (Carr et al. 2018), and contextual fear conditioning tests (Villasana et al. 2010, 2013, Cherry et al. 2012, Raber et al. 2013, 2014). Many studies were performed with murine models and high-energy ^{56}Fe ions (500 or 1000 MeV/n) which are a large contributor to the effective dose accumulated during missions (Durante and Cucinotta 2008). Studies combining heavy ions with protons started recently (Kiffer et al. 2018). In order to come closer to a chronic exposure scenario than with one acute irradiation, fractionated exposures with iron ions were performed, revealing that fractionated ^{56}Fe ion exposure was similarly detrimental for adult neural stem cells as acute exposure (Rivera et al. 2013). The deficits concern memory, object recognition, learning processes, orientation in space, motor function, reaction time, and can also express themselves as depression-like behavior and fatigue (Shukitt-Hale et al. 2000, 2007, Higuchi et al. 2002, Villasana et al. 2010, Britten et al. 2012, 2016, Cherry et al. 2012, Raber et al. 2014, Parihar et al. 2016, 2018, Wyrobek and Britten 2016). Such deficits occur also during the cognitive decline observed in aged animals. The effects are observed in rodents in the dose range relevant to ISS (100–200 mSv) (Cherry et al. 2012, Britten et al. 2014, Wyrobek and Britten 2016, Carr et al. 2018, Kiffer et al. 2019) and exploration missions and persist for 1 year after exposure (Parihar et al. 2018). In rats, social processing was affected after as low doses as 10 and 100 mGy of oxygen ions (1000 MeV/n) (Jones et al. 2019). No effects on locomotor activity of iron ion exposed male Sprague-Dawley rats and male and female C57BL/6 mice were observed 3 months after irradiation at doses of 1–4 Gy (Casadesus et al. 2004, Villasana et al. 2010).

Possible functional and structural changes in the brain after heavy ion exposure include oxidative stress, genomic instability, neuroinflammation, degradation of neuronal architecture including modulated dendritic spine morphology and density, e.g., in the dentate gyrus and cornu Ammon 1 within the hippocampus, suppression of hippocampal neurogenesis, and synaptic integrity with altered pre- and postsynaptic gene expression and neurotransmitter release (Machida et al. 2010, DeCarolis et al. 2014, Allen et al. 2015, Impey et al. 2016, Raber et al. 2016, Sweet et al. 2016, Acharya et al. 2017, Carr et al. 2018, Dickstein et al. 2018). Electrophysiological changes in hippocampal slices of proton-irradiated mice were observed at doses of 1 Gy (Marty et al. 2014). The connection between radiation exposure and oxidative stress is discussed since decades (Cucinotta and Durante 2009), and oxidative stress is believed to play a role in neurodegenerative disease (Huang et al. 2016, Tenkorang et al. 2018), suggesting that ionizing radiation could initiate or promote neurodegeneration via induction of oxidative stress.

Particularly pronounced effects are expected for heavy ions with atomic numbers greater than 10 ($Z > 10$), so NASA has introduced an additional threshold for exposure of the CNS to these ions (Table 3.2) (Williams 2015). The extent to which astronauts are at risk for cognitive deficits and for the earlier onset of dementia is unclear.

Table 3.2 Limits for short-term effects and non-cancer effects for space missions during the whole career, NASA Standard 3001 (Williams 2015)

Organ	30-day limit	1-year limit	Career limit
Eye lens[a]	1000 mGy-Eq[d]	2000 mGy-Eq	4000 mGy-Eq
Skin	1500 mGy-Eq	3000 mGy-Eq	6000 mGy-Eq
Blood-forming organs	250 mGy-Eq	500 mGy-Eq	n/a
Cardiovascular system[b]	250 mGy-Eq	500 mGy-Eq	1000 mGy-Eq
Central nervous system (CNS)[c]	500 mGy	1000 mGy	1500 mGy
CNS[c] (Z>10)		100 mGy	250 mGy

[a]These limits prevent cataract formation within 5 years. Cataracts induced by small GCR doses cannot be prevented at present

[b]Average limits for myocardium and adjacent arteries

[c]These limits were calculated for the hippocampus

[d]mGy-Eq, milliGray-equivalent—This dose is based on the *relative biological effectiveness*, RBE which is specific for the effect and the radiation quality, and the energy dose. For non-cancer effects on eye lens, skin, blood-forming organs and the cardiovascular system, an RBE of 6 for neutrons (1–5 MeV) or 3.5 (5–50 MeV), 2.5 for heavy ions, and 1.5 for protons with an energy above 2 MeV were recommended. The RBE for CNS effects is largely unknown; therefore, the limit is given as energy dose in mGy, with an additional limit for ions with an atomic number >10

Furthermore, recent studies revealed sex differences in the neurobehavioral changes after heavy ions. In the dose range below 1 Gy, female rodents had less severe deficits than males, which was partly ascribed to neuroprotective actions of estrogen (Silasi et al. 2004, Whoolery et al. 2017, Krukowski et al. 2018). The microglia of female mice appeared to be less aggressive and thereby contributed less to the heavy ion-induced brain damage than in male mice (Krukowski et al. 2018). At doses above 1 Gy γ-rays or ^{56}Fe ions, female mice were more susceptible than male mice (Villasana et al. 2006, 2011). This underlines the necessity to perform studies using spaceflight relevant (not to high) dose ranges.

Studies on predispositions for radiation-induced cognitive deficits are still in an early stage. Not all iron ion (1 GeV/n ^{56}Fe) exposed rats develop cognitive impairments even at higher doses (up to 2 Gy), and a hippocampal proteome analysis revealed a specific protein expression profile in susceptible rats which is suggested to promote neuronal loss and apoptosis and to change synaptic plasticity and dendritic remodeling (Britten et al. 2017, Dutta et al. 2018). Specific alleles of apolipoprotein E[2] (ApoE), resulting in expression of the ApoE4 isoform, represent a major genetic risk factor for Alzheimer's Disease (AD). Radiation exposure reinforced genotype-dependent neurobehavioral deficits in transgenic mice expressing

---

[2]ApoE is present in serum. As recognition site for low-density lipoprotein (LDL) receptors, it is involved in transport, deposit, and metabolism of cholesterol within very low-density lipoproteins and chylomicrons [Blum, C. B., R. J. Deckelbaum, L. D. Witte, A. R. Tall and J. Cornicelli (1982). "Role of apolipoprotein E-containing lipoproteins in abetalipoproteinemia." J Clin Invest 70(6): 1157–1169, Blum, C. B. (2016). "Type III Hyperlipoproteinemia: Still Worth Considering?" Prog Cardiovasc Dis 59(2): 119–124.]. In humans, the codons for the amino acid positions 112 and 158 are polymorph (E2 cysteine/cysteine; E3 cysteine/arginine; E4 arginine/arginine), resulting in six different genotypes (ε2/ε2, ε3/ε3, ε4/ε4; ε2/ε3, ε2/ε4, ε3/ε4).

different human ApoE isoforms (Villasana et al. 2006, 2011, 2016, Haley et al. 2012, Yeiser et al. 2013). Chronic low-dose γ-rays exposure (1 or 20 mGy/day) over 300 days changed the phosphoproteome in the hippocampus of ApoE$^{-/-}$ mice (Kempf et al. 2016). Also, mutations in presenilin and amyloid precursors might contribute to the development of AD. In murine AD models with these mutations, heavy ion exposure can expedite the disease process and behavioral deficits (Vlkolinsky et al. 2010, Cherry et al. 2012).

3.5.2.3 Effects on the Cardiovascular System

Evidence for radiation-induced cardiovascular pathologies such as atherosclerosis and cerebrovascular disease stems from studies with radiotherapy patients with fractionated partial body exposure and atomic bomb survivors who experienced an acute whole body exposure.

Radiation-induced cardiovascular diseases including accelerated atherosclerosis, myocardial fibrosis, and cardiac conduction and valve abnormalities were observed as side effect of cancer radiotherapy, sometimes years or decades after radiation exposure (Darby et al. 2005, 2013, McGale et al. 2011). In these cases, some blood vessels or even the heart were exposed to high radiation doses, as, due to technical limitations, large volumes of non-cancer (normal) tissues surround the tumors were irradiated (Boerma 2016). DNA damage, ROS, and chronic inflammation are suggested to contribute to the development of cardiovascular diseases such as atherosclerosis following exposure to therapeutic doses of ionization radiation (Sylvester et al. 2018).

Evidence for the contribution of ionizing radiation exposure to the development of cardiovascular diseases after acute whole body exposure comes from epidemiological studies with the atomic bomb survivors and nuclear workers (Little et al. 2012). From these studies, a risk for radiation-induced cardiovascular diseases at doses above 0.5 Gy of low-LET radiation was derived (Little et al. 2012, Little 2016, Sylvester et al. 2018). This dose could be exceeded, e.g., by a total stay on the ISS in LEO for ~3–5 years (depending on solar activity) (which has never happened up to now) or by a 3-year Mars mission. Therefore, possible degenerative tissue effects following exposures to GCR or SPEs expected during long-duration space-flight cannot be excluded at the current stage of knowledge.

Currently, the cardiovascular disease risk from chronic exposure to GCR is derived from data from atomic bomb survivors in Japan (Shimizu et al. 2010) and meta-analyzes of occupational or environmental exposures (Little et al. 2012, Little 2016). The relative risk for all types of cardiovascular disease was estimated to be 1.10–1.20 at a dose of 1 Sv (Little et al. 2012, Little 2016). A publication on allegedly increased cardiovascular mortality among Apollo astronauts (Delp et al. 2016) sparked controversy and was considered untenable due to deficiencies in data collection and analysis (Cucinotta et al. 2016). A cohort-based reanalysis considering the different age structure of Apollo and nonlunar astronauts failed to demonstrate an excess cardiovascular disease-caused death in lunar astronauts (Reynolds and

Day 2017). In the US astronaut population, cardiovascular disease mortality was significantly lower than in the US population, as explained by the rigorous astronaut selection and healthy lifestyle of the astronauts (Cucinotta et al. 2016). This healthy worker effect is confirmed by the recent mortality study by Reynolds and Day (2017). Based on a recent study including 310 NASA astronauts and 981 nonastronaut NASA employees, a similar risk for cardiovascular disease was found (Ade et al. 2017). These results underline the difficulties of epidemiological studies on long-term effects of space travel, as the astronaut cohort is small and large confounding effects meet quite low doses of radiation that astronauts were exposed to (Reynolds and Day 2017, Sylvester et al. 2018). A multi-year Mars mission will exceed the 0.5 Gy threshold for radiation-induced cardiovascular disease that was derived from studies with radiation-exposed terrestrial cohorts. The current NASA limits for the circulatory system are listed in Table 3.2.

The role of radiation quality and dose-rate effects in radiation-induced cardiovascular disease is an important prerequisite for extrapolation from Earth-based epidemiological data to the exposure situation during space missions. Therefore, the effects of high-LET radiation on the cardiovascular system are studied. These studies can also improve the understanding of mechanisms and pathways leading to radiation-induced cardiovascular disease. Ground-based studies with animal, mostly rodent, and cell culture models are performed at heavy ion accelerators. Many mouse strains are resistant to atherosclerosis[3]; therefore, as in other atherosclerosis studies, mutant mice with defects in the lipid metabolism[4] are used to study radiation-induced atherosclerosis (Sylvester et al. 2018). These mice are predisposed to develop atherosclerotic lesions and the time until they are observable is reduced compared to wild-type mice (Meir and Leitersdorf 2004, Sylvester et al. 2018), which might be explained by different inflammatory and thrombotic pathway activation in wild-type and in ApoE-deficient (ApoE$^{-/-}$) mice (Hoving et al. 2012). After exposure to ^{56}Fe ions (600 MeV/n, 2.6 Gy), vascular reactivity was decreased at 4–5 weeks post-irradiation, before atherosclerotic plaques developed (White et al. 2015), and 2 and 5 Gy ^{56}Fe ions accelerated the development of atherosclerotic lesions in exposed arteries in ApoE$^{-/-}$ mice (Yu et al. 2011). Low-dose ^{56}Fe ions (1 GeV/n, 150 mGy) irradiation modulated several pathways in cardiomyocytes including inflammation, immune cell trafficking, DNA damage, and repair and free-radical scavenging (Coleman et al. 2015). In wild-type mice, 100 mGy protons (150 MeV) had no effect on actin, collagen type III and inflammatory cell markers

---

[3] This resistance is explained by their short life span and high concentration of the protective high-density lipoprotein (HDL) and a low concentration of low-density lipoprotein (LDL) in plasma.

[4] Apolipoprotein E (ApoE)- and LDL-receptor (LDLr)-knockouts, ApoE3 Leiden- and ApoB-100-mutants acting in a dominant negative fashion [Emini Veseli, B., P. Perrotta, G. R. A. De Meyer, L. Roth, C. Van der Donckt, W. Martinet and G. R. Y. De Meyer (2017). "Animal models of atherosclerosis." Eur J Pharmacol 816: 3–13. & Getz, G. S. and C. A. Reardon (2015). "Use of Mouse Models in Atherosclerosis Research." Methods Mol Biol 1339: 1–16.]. It has to be considered that plaque location, form, structure, and stability in these models differs from those in human atherosclerosis [Getz, G. S. and C. A. Reardon (2012). "Animal models of atherosclerosis." Arterioscler Thromb Vasc Biol 32(5): 1104–1115.]

expression in the heart, while 500 mGy ^{56}Fe ions (600 MeV/n) had a strong effect and this was prevented by prior proton exposure (Ramadan et al. 2016). In rats, γ-ray, proton, and heavy ion exposure reduced the plasma level of total antioxidants in Sprague-Dawley rats (Guan et al. 2004). Whole body exposure to 1 Gy iron ions increased aortic stiffness in rats, and ROS production by radiation-induced xanthine oxidase activation was suggested as underlying mechanism (Soucy et al. 2011).

NASA identified several gaps in the knowledge on radiation risk for cardiovascular disease as long-term health effect for astronauts on exploration missions characterize and its mitigation strategy includes biological risk characterization, risk assessment, and medical countermeasures (Patel et al. 2016).

References

Acharya MM, Baddour AA, Kawashita T, Allen BD, Syage AR, Nguyen TH, Yoon N, Giedzinski E, Yu L, Parihar VK, Baulch JE (2017) Epigenetic determinants of space radiation-induced cognitive dysfunction. Sci Rep 7:42885

Ade CJ, Broxterman RM, Charvat JM, Barstow TJ (2017) Incidence rate of cardiovascular disease end points in the national aeronautics and space administration astronaut corps. J Am Heart Assoc 6(8):e005564

Ainsworth EJ, Kelly LS, Mahlmann LJ, Schooley JC, Thomas RH, Howard J, Alpen EL (1983) Response of colony-forming units-spleen to heavy charged particles. Radiat Res 96(1):180–197

Allen AR, Raber J, Chakraborti A, Sharma S, Fike JR (2015) (56)Fe irradiation alters spine density and dendritic complexity in the mouse hippocampus. Radiat Res 184(6):586–594

Alpen EL, Powers-Risius P, Curtis SB, DeGuzman R (1993) Tumorigenic potential of high-Z, high-LET charged-particle radiations. Radiat Res 136(3):382–391

Alpen EL, Powers-Risius P, Curtis SB, DeGuzman R, Fry RJ (1994) Fluence-based relative biological effectiveness for charged particle carcinogenesis in mouse Harderian gland. Adv Space Res 14(10):573–581

Ando K, Koike S, Oohira C, Ogiu T, Yatagai F (2005) Tumor induction in mice locally irradiated with carbon ions: a retrospective analysis. J Radiat Res 46(2):185–190

Ando K, Koike S, Ohmachi Y, Ando Y, Kobashi G (2014) Tumor induction in mice after local irradiation with single doses of either carbon-ion beams or gamma rays. Int J Radiat Biol 90(12):1119–1124

Aoki M, Furusawa Y, Yamada T (2000) LET dependency of heavy-ion induced apoptosis in V79 cells. J Radiat Res 41(2):163–175

Asaithamby A, Chen DJ (2011) Mechanism of cluster DNA damage repair in response to high-atomic number and energy particles radiation. Mutat Res 711(1–2):87–99

Asaithamby A, Uematsu N, Chatterjee A, Story MD, Burma S, Chen DJ (2008) Repair of HZE-particle-induced DNA double-strand breaks in normal human fibroblasts. Radiat Res 169(4):437–446

Banath JP, Olive PL (2003) Expression of phosphorylated histone H2AX as a surrogate of cell killing by drugs that create DNA double-strand breaks. Cancer Res 63(15):4347–4350

Barcellos-Hoff MH, Mao JH (2016) HZE radiation non-targeted effects on the microenvironment that mediate mammary carcinogenesis. Front Oncol 6:57

Barendsen GW, Beusker TL, Vergroesen AJ, Budke L (1960) Effects of different radiations on human cells in tissue culture. II. Biological experiments. Radiat Res 13:841–849

Barendsen GW, Walter HM, Fowler JF, Bewley DK (1963) Effects of different ionizing radiations on human cells in tissue culture. III. Experiments with cyclotron-accelerated alpha-particles and deuterons. Radiat Res 18:106–119

Baskar R (2010) Emerging role of radiation induced bystander effects: cell communications and carcinogenesis. Genome Integr 1(1):13

Baulch JE, Craver BM, Tran KK, Yu L, Chmielewski N, Allen BD, Limoli CL (2015) Persistent oxidative stress in human neural stem cells exposed to low fluences of charged particles. Redox Biol 5:24–32

Baumstark-Khan C, Heilmann J, Rink H (2003) Induction and repair of DNA strand breaks in bovine lens epithelial cells after high LET irradiation. Adv Space Res 31(6):1583–1591

Bekker-Jensen S, Mailand N (2010) Assembly and function of DNA double-strand break repair foci in mammalian cells. DNA Repair (Amst) 9(12):1219–1228

Belli M, Cera F, Cherubini R, Ianzini F, Moschini G, Sapora O, Simone G, Tabocchini MA, Tiveron P (1991) Mutation induction and RBE-LET relationship of low-energy protons in V79 cells. Int J Radiat Biol 59(2):459–465

Belli M, Ianzini F, Sapora O, Tabocchini MA, Cera F, Cherubini R, Haque AM, Moschini G, Tiveron P, Simone G (1996) DNA double strand break production and rejoining in V79 cells irradiated with light ions. Adv Space Res 18(1–2):73–82

Belli M, Cherubini R, Dalla Vecchia M, Dini V, Esposito G, Moschini G, Sapora O, Signoretti C, Simone G, Sorrentino E, Tabocchini MA (2001) DNA fragmentation in mammalian cells exposed to various light ions. Adv Space Res 27(2):393–399

Bettega D, Calzolari P, Hessel P, Stucchi CG, Weyrather WK (2009) Neoplastic transformation induced by carbon ions. Int J Radiat Oncol Biol Phys 73(3):861–868

Blakely E, Chang P, Lommel L, Bjornstad K, Dixon M, Tobias C, Kumar K, Blakely WF (1989) Cell-cycle radiation response: role of intracellular factors. Adv Space Res 9(10):177–186

Blakely EA, Daftari IK, Meecham WJ, Alonso LC, Collier JM, Kroll SM, Gillette EL, Lee AC, Lett JT, Cox AB, Castro JR, Char DH (1994) Helium-ion-induced human cataractogenesis. Adv Space Res 14(10):501–505

Blakely EA, Kleiman NJ, Neriishi K, Chodick G, Chylack LT, Cucinotta FA, Minamoto A, Nakashima E, Kumagami T, Kitaoka T, Kanamoto T, Kiuchi Y, Chang P, Fujii N, Shore RE (2010) Radiation cataractogenesis: epidemiology and biology. Radiat Res 173(5):709–717

Boerma M (2016) An introduction to space radiation and its effects on the cardiovascular system: 12

Brenner DJ, Ward JF (1992) Constraints on energy deposition and target size of multiply damaged sites associated with DNA double-strand breaks. Int J Radiat Biol 61(6):737–748

Brenner DJ, Medvedovsky C, Huang Y, Worgul BV (1993) Accelerated heavy particles and the lens. VIII. Comparisons between the effects of acute low doses of iron ions (190 keV/microns) and argon ions (88 keV/microns). Radiat Res 133(2):198–203

Britten RA, Davis LK, Johnson AM, Keeney S, Siegel A, Sanford LD, Singletary SJ, Lonart G (2012) Low (20 cGy) doses of 1 GeV/u (56)Fe--particle radiation lead to a persistent reduction in the spatial learning ability of rats. Radiat Res 177(2):146–151

Britten RA, Davis LK, Jewell JS, Miller VD, Hadley MM, Sanford LD, Machida M, Lonart G (2014) Exposure to mission relevant doses of 1 GeV/nucleon (56)Fe particles leads to impairment of attentional set-shifting performance in socially mature rats. Radiat Res 182(3):292–298

Britten RA, Jewell JS, Miller VD, Davis LK, Hadley MM, Wyrobek AJ (2016) Impaired spatial memory performance in adult wistar rats exposed to low (5-20 cGy) doses of 1 GeV/n (56)Fe particles. Radiat Res 185(3):332–337

Britten RA, Jewell JS, Davis LK, Miller VD, Hadley MM, Semmes OJ, Lonart G, Dutta SM (2017) Changes in the hippocampal proteome associated with spatial memory impairment after exposure to low (20 cGy) doses of 1 GeV/n (56)Fe radiation. Radiat Res 187(3):287–297

Brooks A, Bao S, Rithidech K, Couch LA, Braby LA (2001) Relative effectiveness of HZE iron-56 particles for the induction of cytogenetic damage in vivo. Radiat Res 155(2):353–359

Bücker H, Facius R, Horneck G, Reitz G, Graul EH, Berger H, Hoffken H, Ruther W, Heinrich W, Beaujean R, Enge W (1986a) Embryogenesis and organogenesis of *Carausius morosus* under spaceflight conditions. Adv Space Res 6(12):115–124

Bücker H, Horneck G, Reitz G, Graul EH, Berger H, Hoffken H, Ruther W, Heinrich W, Beaujean R (1986b) Embryogenesis and organogenesis of *Carausius morosus* under spaceflight conditions. Naturwissenschaften 73(7):433–434

Carr H, Alexander TC, Groves T, Kiffer F, Wang J, Price E, Boerma M, Allen AR (2018) Early effects of (16)O radiation on neuronal morphology and cognition in a murine model. Life Sci Space Res (Amst) 17:63–73

Casadesus G, Shukitt-Hale B, Cantuti-Castelvetri I, Rabin BM, Joseph JA (2004) The effects of heavy particle irradiation on exploration and response to environmental change. Adv Space Res 33(8):1340–1346

Chang J, Feng W, Wang Y, Luo Y, Allen AR, Koturbash I, Turner J, Stewart B, Raber J, Hauer-Jensen M, Zhou D, Shao L (2015) Whole-body proton irradiation causes long-term damage to hematopoietic stem cells in mice. Radiat Res 183(2):240–248

Chao NJ (2007) Accidental or intentional exposure to ionizing radiation: biodosimetry and treatment options. Exp Hematol 35(4 Suppl 1):24–27

Chen DJ, Strniste GF, Tokita N (1984) The genotoxicity of alpha particles in human embryonic skin fibroblasts. Radiat Res 100(2):321–327

Cherry JD, Liu B, Frost JL, Lemere CA, Williams JP, Olschowka JA, O'Banion MK (2012) Galactic cosmic radiation leads to cognitive impairment and increased abeta plaque accumulation in a mouse model of Alzheimer's disease. PLoS One 7(12):e53275

Chishti AA, Baumstark-Khan C, Koch K, Kolanus W, Feles S, Konda B, Azhar A, Spitta LF, Henschenmacher B, Diegeler S, Schmitz C, Hellweg CE (2018) Linear energy transfer modulates radiation-induced NF-kappa B activation and expression of its downstream target genes. Radiat Res 189(4):354–374

Coleman MA, Sasi SP, Onufrak J, Natarajan M, Manickam K, Schwab J, Muralidharan S, Peterson LE, Alekseyev YO, Yan X, Goukassian DA (2015) Low-dose radiation affects cardiac physiology: gene networks and molecular signaling in cardiomyocytes. Am J Physiol Heart Circ Physiol 309(11):H1947–H1963

Committee on the Longitudinal Study of Astronaut Health (2004) Review of NASA's longitudinal study of astronaut health (LSAH). In: Longnecker DE, Manning FJ, Worth MH (ed). The National Academic Press, Washington, DC

Coppe JP, Desprez PY, Krtolica A, Campisi J (2010) The senescence-associated secretory phenotype: the dark side of tumor suppression. Annu Rev Pathol 5:99–118

Cornelissen M, Thierens H, De Ridder L (2002) Interphase death in human peripheral blood lymphocytes after moderate and high doses of low and high LET radiation: an electron microscopic approach. Anticancer Res 22(1a):241–245

Cornforth MN (2006) Perspectives on the formation of radiation-induced exchange aberrations. DNA Repair (Amst) 5(9–10):1182–1191

Costes SV, Boissiere A, Ravani S, Romano R, Parvin B, Barcellos-Hoff MH (2006) Imaging features that discriminate between foci induced by high- and low-LET radiation in human fibroblasts. Radiat Res 165(5):505–515

Cox AB, Lee AC, Williams GR, Lett JT (1992) Late cataractogenesis in primates and lagomorphs after exposure to particulate radiations. Adv Space Res 12(2–3):379–384

Cronkite EP (1964) The diagnosis, treatment, and prognosis of human radiation injury from whole-body exposure. Ann NY Acad Sci 114:341–355

Cucinotta FA, Chappell LJ (2011) Updates to astronaut radiation limits: radiation risks for never-smokers. Radiat Res 176(1):102–114

Cucinotta FA, Durante M (2006) Cancer risk from exposure to galactic cosmic rays: implications for space exploration by human beings. Lancet Oncol 7(5):431–435

Cucinotta FA, Durante M (2009) Risk of radiation carcinogenesis. In: Mcphee J, Charles J (eds) Human health and performance risks of space exploration missions. National Aeronautics and Space Administration, Houston, TX, pp 119–170

Cucinotta FA, Nikjoo H, Goodhead DT (2000) Model for radial dependence of frequency distributions for energy imparted in nanometer volumes from HZE particles. Radiat Res 153(4):459–468

Cucinotta FA, Manuel FK, Jones J, Iszard G, Murrey J, Djojonegro B, Wear M (2001a) Space radiation and cataracts in astronauts. Radiat Res 156(5 Pt 1):460–466

Cucinotta FA, Schimmerling W, Wilson JW, Peterson LE, Badhwar GD, Saganti PB, Dicello JF (2001b) Space radiation cancer risks and uncertainties for Mars missions. Radiat Res 156(5 Pt 2):682–688

Cucinotta FA, Kim M-HY, Ren L (2005) Managing lunar and mars mission radiation risks. Part 1; cancer risks, uncertainties, and shielding effectiveness. Technical Publication, NASA Johnson Space Center, Houston, TX, p 44

Cucinotta FA, Alp M, Rowedder B, Kim MH (2015) Safe days in space with acceptable uncertainty from space radiation exposure. Life Sci Space Res (Amst) 5:31–38

Cucinotta FA, Hamada N, Little MP (2016) No evidence for an increase in circulatory disease mortality in astronauts following space radiation exposures. Life Sci Space Res (Amst) 10:53–56

Cucinotta FA, To K, Cacao E (2017) Predictions of space radiation fatality risk for exploration missions. Life Sci Space Res (Amst) 13:1–11

Cucinotta FA, Cacao E, Kim MY, Saganti PB (2019) Non-targeted effects lead to a paridigm shift in risk assessment for a mission to the earth's moon or martian moon phobos. Radiat Prot Dosim 183(1–2):213–218

Dainiak N, Waselenko JK, Armitage JO, MacVittie TJ, Farese AM (2003) The hematologist and radiation casualties. Hematol Am Soc Hematol Educ Program 2003:473–496

Darby SC, McGale P, Taylor CW, Peto R (2005) Long-term mortality from heart disease and lung cancer after radiotherapy for early breast cancer: prospective cohort study of about 300,000 women in US SEER cancer registries. Lancet Oncol 6(8):557–565

Darby SC, Ewertz M, McGale P, Bennet AM, Blom-Goldman U, Bronnum D, Correa C, Cutter D, Gagliardi G, Gigante B, Jensen MB, Nisbet A, Peto R, Rahimi K, Taylor C, Hall P (2013) Risk of ischemic heart disease in women after radiotherapy for breast cancer. N Engl J Med 368(11):987–998

Datta K, Suman S, Kallakury BV, Fornace AJ Jr (2012) Exposure to heavy ion radiation induces persistent oxidative stress in mouse intestine. PLoS One 7(8):e42224

DeCarolis NA, Rivera PD, Ahn F, Amaral WZ, LeBlanc JA, Malhotra S, Shih HY, Petrik D, Melvin N, Chen BP, Eisch AJ (2014) (56)Fe particle exposure results in a long-lasting increase in a cellular index of genomic instability and transiently suppresses adult hippocampal neurogenesis in vivo. Life Sci Space Res (Amst) 2:70–79

DeFazio LG, Stansel RM, Griffith JD, Chu G (2002) Synapsis of DNA ends by DNA-dependent protein kinase. EMBO J 21(12):3192–3200

Delp MD, Charvat JM, Limoli CL, Globus RK, Ghosh P (2016) Apollo lunar astronauts show higher cardiovascular disease mortality: possible deep space radiation effects on the vascular endothelium. Sci Rep 6:29901

Denisova NA, Shukitt-Hale B, Rabin BM, Joseph JA (2002) Brain signaling and behavioral responses induced by exposure to (56)Fe-particle radiation. Radiat Res 158(6):725–734

Deorukhkar A, Krishnan S (2010) Targeting inflammatory pathways for tumor radiosensitization. Biochem Pharmacol 80(12):1904–1914

Desai N, Davis E, O'Neill P, Durante M, Cucinotta FA, Wu H (2005) Immunofluorescence detection of clustered gamma-H2AX foci induced by HZE-particle radiation. Radiat Res 164(4 Pt 2):518–522

Di Leonardo A, Linke SP, Clarkin K, Wahl GM (1994) DNA damage triggers a prolonged p53-dependent G1 arrest and long-term induction of Cip1 in normal human fibroblasts. Genes Dev 8(21):2540–2551

Dicello JF, Christian A, Cucinotta FA, Gridley DS, Kathirithamby R, Mann J, Markham AR, Moyers MF, Novak GR, Piantadosi S, Ricart-Arbona R, Simonson DM, Strandberg JD, Vazquez M, Williams JR, Zhang Y, Zhou H, Huso D (2004) In vivo mammary tumourigenesis in the Sprague-Dawley rat and microdosimetric correlates. Phys Med Biol 49(16):3817–3830

Dickstein DL, Talty R, Bresnahan E, Varghese M, Perry B, Janssen WGM, Sowa A, Giedzinski E, Apodaca L, Baulch J, Acharya M, Parihar V, Limoli CL (2018) Alterations in synaptic density and myelination in response to exposure to high-energy charged particles. J Comp Neurol 526(17):2845–2855

Donnelly EH, Nemhauser JB, Smith JM, Kazzi ZN, Farfan EB, Chang AS, Naeem SF (2010) Acute radiation syndrome: assessment and management. South Med J 103(6):541–546

Dorr H, Meineke V (2011) Acute radiation syndrome caused by accidental radiation exposure— therapeutic principles. BMC Med 9:126

Drouet M, Herodin F (2010) Radiation victim management and the haematologist in the future: time to revisit therapeutic guidelines? Int J Radiat Biol 86(8):636–648

Durante M, Cucinotta FA (2008) Heavy ion carcinogenesis and human space exploration. Nat Rev Cancer 8(6):465–472

Durante M, Bonassi S, George K, Cucinotta FA (2001) Risk estimation based on chromosomal aberrations induced by radiation. Radiat Res 156(5 Pt 2):662–667

Dutta SM, Hadley MM, Peterman S, Jewell JS, Duncan VD, Britten RA (2018) Quantitative proteomic analysis of the hippocampus of rats with GCR-induced spatial memory impairment. Radiat Res 189(2):136–145

Eccles LJ, Lomax ME, O'Neill P (2010) Hierarchy of lesion processing governs the repair, double-strand break formation and mutability of three-lesion clustered DNA damage. Nucleic Acids Res 38(4):1123–1134

Eriksson D, Stigbrand T (2010) Radiation-induced cell death mechanisms. Tumour Biol 31(4):363–372

Fei P, El-Deiry WS (2003) P53 and radiation responses. Oncogene 22(37):5774–5783

Fliedner TM, Andrews GA, Cronkite EP, Bond VP (1964) Early and late cytologic effects of whole body irradiation on human marrow. Blood 23:471–487

Fournier C, Taucher-Scholz G (2004) Radiation induced cell cycle arrest: an overview of specific effects following high-LET exposure. Radiother Oncol 73(Suppl 2):S119–S122

Freund A, Patil CK, Campisi J (2011) p38MAPK is a novel DNA damage response-independent regulator of the senescence-associated secretory phenotype. EMBO J 30(8):1536–1548

Fry RJM, Storer JB (1987) External radiation carcinogenesis. Adv Radiat Biol 13:31–90. J. T. Lett, Elsevier

Fry RJ, Powers-Risius P, Alpen EL, Ainsworth EJ, Ullrich RL (1983) High-LET radiation carcinogenesis. Adv Space Res 3(8):241–248

Fry RJ, Powers-Risius P, Alpen EL, Ainsworth EJ (1985) High-LET radiation carcinogenesis. Radiat Res Suppl 8:S188–S195

Gaugler MH, Vereycken-Holler V, Squiban C, Vandamme M, Vozenin-Brotons MC, Benderitter M (2005) Pravastatin limits endothelial activation after irradiation and decreases the resulting inflammatory and thrombotic responses. Radiat Res 163(5):479–487

George K, Durante M, Wu H, Willingham V, Badhwar G, Cucinotta FA (2001) Chromosome aberrations in the blood lymphocytes of astronauts after space flight. Radiat Res 156(6):731–738

George K, Durante M, Willingham V, Wu H, Yang TC, Cucinotta FA (2003) Biological effectiveness of accelerated particles for the induction of chromosome damage measured in metaphase and interphase human lymphocytes. Radiat Res 160(4):425–435

George K, Willingham V, Cucinotta FA (2005) Stability of chromosome aberrations in the blood lymphocytes of astronauts measured after space flight by FISH chromosome painting. Radiat Res 164(4 Pt 2):474–480

Ghosh S, Hayden MS (2008) New regulators of NF-kappaB in inflammation. Nat Rev Immunol 8(11):837–848

Goodhead DT (1988) Spatial and temporal distribution of energy. Health Phys 55(2):231–240

Goodhead DT, Thacker J, Cox R (1993) Weiss lecture. Effects of radiations of different qualities on cells: molecular mechanisms of damage and repair. Int J Radiat Biol 63(5):543–556

Gourmelon P, Marquette C, Agay D, Mathieu J, Clarencon D (2005) Involvement of the central nervous system in radiation-induced multi-organ dysfunction and/or failure. BJR Suppl 27:62–68

Gridley DS, Coutrakon GB, Rizvi A, Bayeta EJ, Luo-Owen X, Makinde AY, Baqai F, Koss P, Slater JM, Pecaut MJ (2008) Low-dose photons modify liver response to simulated solar particle event protons. Radiat Res 169(3):280–287

Guan J, Wan XS, Zhou Z, Ware J, Donahue JJ, Biaglow JE, Kennedy AR (2004) Effects of dietary supplements on space radiation-induced oxidative stress in Sprague-Dawley rats. Radiat Res 162(5):572–579

Gudkov AV, Komarova EA (2010a) Pathologies associated with the p53 response. Cold Spring Harb Perspect Biol 2(7):a001180

Gudkov AV, Komarova EA (2010b) Radioprotection: smart games with death. J Clin Invest 120(7):2270–2273

Gueguen Y, Bontemps A, Ebrahimian TG (2019) Adaptive responses to low doses of radiation or chemicals: their cellular and molecular mechanisms. Cell Mol Life Sci 76(7):1255–1273

Guida P, Vazquez ME, Otto S (2005) Cytotoxic effects of low- and high-LET radiation on human neuronal progenitor cells: induction of apoptosis and TP53 gene expression. Radiat Res 164(4 Pt 2):545–551

Habelhah H (2010) Emerging complexity of protein ubiquitination in the NF-kappaB pathway. Genes Cancer 1(7):735–747

Haley GE, Villasana L, Dayger C, Davis MJ, Raber J (2012) Apolipoprotein e genotype-dependent paradoxical short-term effects of (56)fe irradiation on the brain. Int J Radiat Oncol Biol Phys 84(3):793–799

Hall EJ, Worgul BV, Smilenov L, Elliston CD, Brenner DJ (2006) The relative biological effectiveness of densely ionizing heavy-ion radiation for inducing ocular cataracts in wild type versus mice heterozygous for the ATM gene. Radiat Environ Biophys 45(2):99–104

Hamada N (2017) Ionizing radiation sensitivity of the ocular lens and its dose rate dependence. Int J Radiat Biol 93(10):1024–1034

Hamada N, Hara T, Funayama T, Sakashita T, Kobayashi Y (2008) Energetic heavy ions accelerate differentiation in the descendants of irradiated normal human diploid fibroblasts. Mutat Res 637(1–2):190–196

Han ZB, Suzuki H, Suzuki F, Suzuki M, Furusawa Y, Kato T, Ikenaga M (1998) Relative biological effectiveness of accelerated heavy ions for induction of morphological transformation in Syrian hamster embryo cells. J Radiat Res 39(3):193–201

Hartwell LH, Weinert TA (1989) Checkpoints: controls that ensure the order of cell cycle events. Science 246(4930):629–634

Heilmann J, Taucher-Scholz G, Kraft G (1995) Induction of DNA double-strand breaks in CHO-K1 cells by carbon ions. Int J Radiat Biol 68(2):153–162

Hellweg CE (2012) Cellular response to exposure with different radiation qualities with a focus on nuclear factor κB habilitation. Freie Universität, Berlin

Hellweg CE (2015) The nuclear factor kappaB pathway: a link to the immune system in the radiation response. Cancer Lett 368(2):275–289

Hellweg CE, Baumstark-Khan C, Schmitz C, Lau P, Meier MM, Testard I, Berger T, Reitz G (2011) Activation of the nuclear factor kappaB pathway by heavy ion beams of different linear energy transfer. Int J Radiat Biol 87(9):954–963

Hellweg CE, Spitta LF, Henschenmacher B, Diegeler S, Baumstark-Khan C (2016) Transcription factors in the cellular response to charged particle exposure. Front Oncol 6:61

Hellweg CE, Spitta LF, Koch K, Chishti AA, Henschenmacher B, Diegeler S, Konda B, Feles S, Schmitz C, Berger T, Baumstark-Khan C (2018) The role of the nuclear factor kappab pathway in the cellular response to low and high linear energy transfer radiation. Int J Mol Sci 19(8):2220

Hendry JH, Potten CS, Merritt A (1995) Apoptosis induced by high- and low-LET radiations. Radiat Environ Biophys 34(1):59–62

Higuchi Y, Nelson GA, Vazquez M, Laskowitz DT, Slater JM, Pearlstein RD (2002) Apolipoprotein E expression and behavioral toxicity of high charge, high energy (HZE) particle radiation. J Radiat Res 43(Suppl):S219–S224

Higurashi M, Conen PE (1973) In vitro chromosomal radiosensitivity in "chromosomal breakage syndromes". Cancer 32(2):380–383

Hirayama R, Ito A, Tomita M, Tsukada T, Yatagai F, Noguchi M, Matsumoto Y, Kase Y, Ando K, Okayasu R, Furusawa Y (2009) Contributions of direct and indirect actions in cell killing by high-LET radiations. Radiat Res 171(2):212–218

Horneck G, Rettberg P, Baumstark-Khan C, Rink H, Kozubek S, Schafer M, Schmitz C (1996) DNA repair in microgravity: studies on bacteria and mammalian cells in the experiments REPAIR and KINETICS. J Biotechnol 47(2–3):99–112

Horneck G, Rettberg P, Kozubek S, Baumstark-Khan C, Rink H, Schafer M, Schmitz C (1997) The influence of microgravity on repair of radiation-induced DNA damage in bacteria and human fibroblasts. Radiat Res 147(3):376–384

Horstmann M, Durante M, Johannes C, Obe G (2005a) Chromosomal intrachanges induced by swift iron ions. Adv Space Res 35(2):276–279

Horstmann M, Durante M, Johannes C, Pieper R, Obe G (2005b) Space radiation does not induce a significant increase of intrachromosomal exchanges in astronauts' lymphocytes. Radiat Environ Biophys 44(3):219–224

Hoving S, Heeneman S, Gijbels MJ, Te Poele JA, Visser N, Cleutjens J, Russell NS, Daemen MJ, Stewart FA (2012) Irradiation induces different inflammatory and thrombotic responses in carotid arteries of wildtype C57BL/6J and atherosclerosis-prone ApoE(−/−) mice. Radiother Oncol 105(3):365–370

Huang L, Snyder AR, Morgan WF (2003) Radiation-induced genomic instability and its implications for radiation carcinogenesis. Oncogene 22(37):5848–5854

Huang WJ, Zhang X, Chen WW (2016) Role of oxidative stress in Alzheimer's disease. Biomed Rep 4(5):519–522

Hwang A, Muschel RJ (1998) Radiation and the G2 phase of the cell cycle. Radiat Res 150(5 Suppl):S52–S59

ICRP (1984) Protection of the public in the event of major radiation accidents—principles for planning. ICRP Publication 40. Ann. ICRP 14 (2). F. D. Sowby, International Commission on Radiological Protection

ICRP (2003) Relative biological effectiveness (RBE), quality factor (Q), and radiation weighting factor (w_R). ICRP Publication 92. Ann. ICRP 33 (4). J. Valentin, International Commission on Radiological Protection

Iliakis G, Wang Y, Guan J, Wang H (2003) DNA damage checkpoint control in cells exposed to ionizing radiation. Oncogene 22(37):5834–5847

Imaoka T, Nishimura M, Kakinuma S, Hatano Y, Ohmachi Y, Yoshinaga S, Kawano A, Maekawa A, Shimada Y (2007) High relative biologic effectiveness of carbon ion radiation on induction of rat mammary carcinoma and its lack of H-ras and Tp53 mutations. Int J Radiat Oncol Biol Phys 69(1):194–203

Impey S, Jopson T, Pelz C, Tafessu A, Fareh F, Zuloaga D, Marzulla T, Riparip LK, Stewart B, Rosi S, Turker MS, Raber J (2016) Short- and long-term effects of (56)Fe irradiation on cognition and hippocampal DNA methylation and gene expression. BMC Genomics 17(1):825

Jackson SP, Bartek J (2009) The DNA-damage response in human biology and disease. Nature 461(7267):1071–1078

Jacob A, Shah KG, Wu R, Wang P (2010) Ghrelin as a novel therapy for radiation combined injury. Mol Med 16(3–4):137–143

Jandial R, Hoshide R, Waters JD, Limoli CL (2018) Space-brain: the negative effects of space exposure on the central nervous system. Surg Neurol Int 9:9

Johannes C, Horstmann M, Durante M, Chudoba I, Obe G (2004) Chromosome intrachanges and interchanges detected by multicolor banding in lymphocytes: searching for clastogen signatures in the human genome. Radiat Res 161(5):540–548

Joiner MC, Marples B, Lambin P, Short SC, Turesson I (2001) Low-dose hypersensitivity: current status and possible mechanisms. Int J Radiat Oncol Biol Phys 49(2):379–389

Jones CB, Mange A, Granata L, Johnson B, Hienz RD, Davis CM (2019) Short and long-term changes in social odor recognition and plasma cytokine levels following oxygen ((16)O) ion radiation exposure. Int J Mol Sci 20(2):339

Jose JG, Ainsworth EJ (1983) Cataract production in mice by heavy charged argon, neon, and carbon particles. Radiat Res 94(3):513–528

Kadhim MA, Macdonald DA, Goodhead DT, Lorimore SA, Marsden SJ, Wright EG (1992) Transmission of chromosomal instability after plutonium alpha-particle irradiation. Nature 355(6362):738–740

Kadhim MA, Hill MA, Moore SR (2006) Genomic instability and the role of radiation quality. Radiat Prot Dosim 122(1–4):221–227

Kadhim M, Salomaa S, Wright E, Hildebrandt G, Belyakov OV, Prise KM, Little MP (2013) Non-targeted effects of ionising radiation—implications for low dose risk. Mutat Res 752(2):84–98

Kajioka EH, Andres ML, Li J, Mao XW, Moyers MF, Nelson GA, Slater JM, Gridley DS (2000) Acute effects of whole-body proton irradiation on the immune system of the mouse. Radiat Res 153(5 Pt 1):587–594

Karagiannis TC, El-Osta A (2004) Double-strand breaks: signaling pathways and repair mechanisms. Cell Mol Life Sci 61(17):2137–2147

Kass EM, Jasin M (2010) Collaboration and competition between DNA double-strand break repair pathways. FEBS Lett 584(17):3703–3708

Kawamura K, Qi F, Kobayashi J (2018) Potential relationship between the biological effects of low-dose irradiation and mitochondrial ROS production. J Radiat Res 59(S2):ii91–ii97

Kempf SJ, Janik D, Barjaktarovic Z, Braga-Tanaka I 3rd, Tanaka S, Neff F, Saran A, Larsen MR, Tapio S (2016) Chronic low-dose-rate ionising radiation affects the hippocampal phosphoproteome in the ApoE$^{-/-}$ Alzheimer's mouse model. Oncotarget 7(44):71817–71832

Kiffer F, Carr H, Groves T, Anderson JE, Alexander T, Wang J, Seawright JW, Sridharan V, Carter G, Boerma M, Allen AR (2018) Effects of (1)H + (16)O charged particle irradiation on short-term memory and hippocampal physiology in a murine model. Radiat Res 189(1):53–63

Khanna KK, Lavin MF, Jackson SP, Mulhern TD (2001) ATM, a central controller of cellular responses to DNA damage. Cell Death Differ 8(11):1052–1065

Kiffer F, Alexander T, Anderson JE, Groves T, Wang J, Sridharan V, Boerma M, Allen AR (2019) Late effects of (16)O-particle radiation on female social and cognitive behavior and hippocampal physiology. Radiat Res 191(3):278–294

Kraft G, Kraft-Weyrather W, Ritter S, Scholz M, Stanton J (1989) Cellular and subcellular effect of heavy ions: a comparison of the induction of strand breaks and chromosomal aberration with the incidence of inactivation and mutation. Adv Space Res 9(10):59–72

Krämer M, Kraft G (1994) Calculations of heavy-ion track structure. Radiat Environ Biophys 33(2):91–109

Krempler A, Deckbar D, Jeggo PA, Lobrich M (2007) An imperfect G2M checkpoint contributes to chromosome instability following irradiation of S and G2 phase cells. Cell Cycle 6(14):1682–1686

Krukowski K, Grue K, Frias ES, Pietrykowski J, Jones T, Nelson G, Rosi S (2018) Female mice are protected from space radiation-induced maladaptive responses. Brain Behav Immun 74:106–120

Kusunoki Y, Hayashi T (2008) Long-lasting alterations of the immune system by ionizing radiation exposure: implications for disease development among atomic bomb survivors. Int J Radiat Biol 84(1):1–14

Law J, Van Baalen M, Foy M, Mason SS, Mendez C, Wear ML, Meyers VE, Alexander D (2014) Relationship between carbon dioxide levels and reported headaches on the international space station. J Occup Environ Med 56(5):477–483

Leach JK, Van Tuyle G, Lin PS, Schmidt-Ullrich R, Mikkelsen RB (2001) Ionizing radiation-induced, mitochondria-dependent generation of reactive oxygen/nitrogen. Cancer Res 61(10):3894–3901

Letaw JR, Silberberg R, Tsao CH (1989) Radiation hazards on space missions outside the magnetosphere. Adv Space Res 9(10):285–291

Lett JT, Cox AB, Keng PC, Lee AC, Su CM, Bergtold DS (1980) Late degeneration in rabbit tissues after irradiation by heavy ions. Life SciSpace Res 18:131–142

Lett JT, Lee AC, Cox AB (1991) Late cataractogenesis in rhesus monkeys irradiated with protons and radiogenic cataract in other species. Radiat Res 126(2):147–156

Lett JT, Lee AC, Cox AB (1994) Risks of radiation cataracts from interplanetary space missions. Acta Astronaut 32(11):739–748

Li M, You L, Xue J, Lu Y (2018) Ionizing radiation-induced cellular senescence in normal, non-transformed cells and the involved DNA damage response: a mini review. Front Pharmacol 9:522

Limoli CL, Giedzinski E, Baure J, Rola R, Fike JR (2007) Redox changes induced in hippocampal precursor cells by heavy ion irradiation. Radiat Environ Biophys 46(2):167–172

Little MP (2016) Radiation and circulatory disease. Mutat Res-Rev Mutat 770:299–318

Little MP, Azizova TV, Bazyka D, Bouffler SD, Cardis E, Chekin S, Chumak VV, Cucinotta FA, de Vathaire F, Hall P, Harrison JD, Hildebrandt G, Ivanov V, Kashcheev VV, Klymenko SV, Kreuzer M, Laurent O, Ozasa K, Schneider T, Tapio S, Taylor AM, Tzoulaki I, Vandoolaeghe WL, Wakeford R, Zablotska LB, Zhang W, Lipshultz SE (2012) Systematic review and meta-analysis of circulatory disease from exposure to low-level ionizing radiation and estimates of potential population mortality risks. Environ Health Perspect 120(11):1503–1511

Löbrich M, Jeggo PA (2007) The impact of a negligent G2/M checkpoint on genomic instability and cancer induction. Nat Rev Cancer 7(11):861–869

Lu C-Y, Ma Y-C, Lin J-M, Chuang C-Y, Sung F-C (2007) Oxidative DNA damage estimated by urinary 8-hydroxydeoxyguanosine and indoor air pollution among non-smoking office employees. Environ Res 103(3):331–337

Lu T, Zhang Y, Kidane Y, Feiveson A, Stodieck L, Karouia F, Ramesh G, Rohde L, Wu H (2017) Cellular responses and gene expression profile changes due to bleomycin-induced DNA damage in human fibroblasts in space. PLoS One 12(3):e0170358

Machida M, Lonart G, Britten RA (2010) Low (60 cGy) doses of (56)Fe HZE-particle radiation lead to a persistent reduction in the glutamatergic readily releasable pool in rat hippocampal synaptosomes. Radiat Res 174(5):618–623

Makale MT, King GL (1993) Plasticity of autonomic control of emesis in irradiated ferrets. AmJPhysiol 265(5 Pt 2):R1092–R1099

Manti L (2006) Does reduced gravity alter cellular response to ionizing radiation? Radiat Environ Biophys 45(1):1–8

Marin A, Martin M, Linan O, Alvarenga F, Lopez M, Fernandez L, Buchser D, Cerezo L (2015) Bystander effects and radiotherapy. Rep Pract Oncol Radiother 20(1):12–21

Marquette C, Linard C, Galonnier M, Van Uye A, Mathieu J, Gourmelon P, Clarençon D (2003) IL-1β, TNFα and IL-6 induction in the rat brain after partial-body irradiation: role of vagal afferents. Int J Radiat Biol 79(10):777–785

Marty VN, Vlkolinsky R, Minassian N, Cohen T, Nelson GA, Spigelman I (2014) Radiation-induced alterations in synaptic neurotransmission of dentate granule cells depend on the dose and species of charged particles. Radiat Res 182(6):653–665

Maxwell CA, Fleisch MC, Costes SV, Erickson AC, Boissiere A, Gupta R, Ravani SA, Parvin B, Barcellos-Hoff MH (2008) Targeted and nontargeted effects of ionizing radiation that impact genomic instability. Cancer Res 68(20):8304–8311

McDonald JT, Kim K, Norris AJ, Vlashi E, Phillips TM, Lagadec C, Della Donna L, Ratikan J, Szelag H, Hlatky L, McBride WH (2010) Ionizing radiation activates the Nrf2 antioxidant response. Cancer Res 70(21):8886–8895

McGale P, Darby SC, Hall P, Adolfsson J, Bengtsson NO, Bennet AM, Fornander T, Gigante B, Jensen MB, Peto R, Rahimi K, Taylor CW, Ewertz M (2011) Incidence of heart disease in 35,000 women treated with radiotherapy for breast cancer in Denmark and Sweden. Radiother Oncol 100(2):167–175

Meir KS, Leitersdorf E (2004) Atherosclerosis in the apolipoprotein-E-deficient mouse: a decade of progress. Arterioscler Thromb Vasc Biol 24(6):1006–1014

Merriam GR Jr, Worgul BV, Medvedovsky C, Zaider M, Rossi HH (1984) Accelerated heavy particles and the lens. I. Cataractogenic potential. Radiat Res 98(1):129–140

Miller RC, Marino SA, Brenner DJ, Martin SG, Richards M, Randers-Pehrson G, Hall EJ (1995) The biological effectiveness of radon-progeny alpha particles. II. Oncogenic transformation as a function of linear energy transfer. Radiat Res 142(1):54–60

Milstead L, Fogarty J, Paloski W (2019) Human research program integrated research plan. National Aeronautics and Space Administration, Houston, TX HRP 47065: 54

Mimitou EP, Symington LS (2009) DNA end resection: many nucleases make light work. DNA Repair (Amst) 8(9):983–995

Moreno-Villanueva M, Wong M, Lu T, Zhang Y, Wu H (2017) Interplay of space radiation and microgravity in DNA damage and DNA damage response. NPJ Microgravity 3:14

Morgan WF (2003) Non-targeted and delayed effects of exposure to ionizing radiation: I. radiation-induced genomic instability and bystander effects in vitro. Radiat Res 159(5):567–580

Mothersill C, Seymour CB (2004) Radiation-induced bystander effects—implications for cancer. Nat Rev Cancer 4(2):158–164

Muller WU, Nusse M, Miller BM, Slavotinek A, Viaggi S, Streffer C (1996) Micronuclei: a biological indicator of radiation damage. Mutat Res 366(2):163–169

Murray D, Mirzayans R, McBride WH (2018) Defenses against pro-oxidant forces—maintenance of cellular and genomic integrity and longevity. Radiat Res 190(4):331–349

Nakanishi M, Niida H, Murakami H, Shimada M (2009) DNA damage responses in skin biology—implications in tumor prevention and aging acceleration. J Dermatol Sci 56(2):76–81

NCRP (1990) NCRP report no. 104—the relative biological effectiveness of radiations of different quality. NCRP, Bethesda, MD, p 98

NCRP (2000) NCRP Report No. 132: Radiation Protection Guidance for Activities in Low-Earth Orbit. NCRP, Bethesda, MD

NCRP (2006) NCRP Report No. 153: Information Needed to Make Radiation Protection Recommendations for Space Missions Beyond Low-Earth Orbit. NCRP, Bethesda, MD, pp 1–427

O'Driscoll M, Jeggo PA (2006) The role of double-strand break repair—insights from human genetics. Nat Rev Genet 7(1):45–54

Obe G, Johannes I, Johannes C, Hallman K, Reitz G, Facius R (1997) Chromosomal aberrations in blood lymphocytes of astronauts after long-term space flights. Int J Radiat Biol 72(6):727–734

Obe G, Facius R, Reitz G, Johannes I, Johannes C (1999) Manned missions to mars and chromosome damage. Int J Radiat Biol 75(4):429–433

Obe G, Pfeiffer P, Savage JR, Johannes C, Goedecke W, Jeppesen P, Natarajan AT, Martinez-Lopez W, Folle GA, Drets ME (2002) Chromosomal aberrations: formation, identification and distribution. Mutat Res 504(1–2):17–36

Parihar VK, Allen BD, Caressi C, Kwok S, Chu E, Tran KK, Chmielewski NN, Giedzinski E, Acharya MM, Britten RA, Baulch JE, Limoli CL (2016) Cosmic radiation exposure and persistent cognitive dysfunction. Sci Rep 6:34774

Parihar VK, Maroso M, Syage A, Allen BD, Angulo MC, Soltesz I, Limoli CL (2018) Persistent nature of alterations in cognition and neuronal circuit excitability after exposure to simulated cosmic radiation in mice. Exp Neurol 305:44–55

Park CC, Henshall-Powell RL, Erickson AC, Talhouk R, Parvin B, Bissell MJ, Barcellos-Hoff MH (2003) Ionizing radiation induces heritable disruption of epithelial cell interactions. Proc Natl Acad Sci USA 100(19):10728–10733

Patel Z, Huff J, Saha J, Wang M, Blattnig S, Wu H (2016) Risk of cardiovascular disease and other degenerative tissue effects from radiation exposure. NASA Human Research Program. Available from: https://humanresearchroadmap.nasa.gov/Risks/risk.aspx?i=98

Pawlik TM, Keyomarsi K (2004) Role of cell cycle in mediating sensitivity to radiotherapy. Int J Radiat Oncol Biol Phys 59(4):928–942

Phillips ER, McKinnon PJ (2007) DNA double-strand break repair and development. Oncogene 26(56):7799–7808

Pierce DA, Preston DL (2000) Radiation-related cancer risks at low doses among atomic bomb survivors. Radiat Res 154(2):178–186

Pierce DA, Shimizu Y, Preston DL, Vaeth M, Mabuchi K (1996) Studies of the mortality of atomic bomb survivors. Report 12, part I. Cancer: 1950–1990. Radiat Res 146(1):1–27

Pierce DA, Shimizu Y, Preston DL, Vaeth M, Mabuchi K (2012) Studies of the mortality of atomic bomb survivors. Report 12, part I. Cancer: 1950–1990. Radiat Res 178(2):Av61–87

Ponomarev AL, Cucinotta FA (2006) Chromatin loops are responsible for higher counts of small DNA fragments induced by high-LET radiation, while chromosomal domains do not affect the fragment sizes. Int J Radiat Biol 82(4):293–305

Preston DL, Ron E, Tokuoka S, Funamoto S, Nishi N, Soda M, Mabuchi K, Kodama K (2007) Solid cancer incidence in atomic bomb survivors: 1958–1998. Radiat Res 168(1):1–64

Prise KM (2006) New advances in radiation biology. Occup Med (Lond) 56(3):156–161

Prise KM, Folkard M, Michael BD (2006) Radiation-induced bystander and adaptive responses in cell and tissue models. Dose Response 4(4):263–276

Purgason A, Zhang Y, Hamilton SR, Gridley DS, Sodipe A, Jejelowo O, Ramesh GT, Moreno-Villanueva M, Wu H (2018) Apoptosis and expression of apoptosis-related genes in mouse intestinal tissue after whole-body proton exposure. Mol Cell Biochem 442(1–2):155–168

Qin S, Schulte BA, Wang GY (2018) Role of senescence induction in cancer treatment. World J Clin Oncol 9(8):180–187

Raber J, Allen AR, Rosi S, Sharma S, Dayger C, Davis MJ, Fike JR (2013) Effects of whole body (56)Fe radiation on contextual freezing and arc-positive cells in the dentate gyrus. Behav Brain Res 246:162–167

Raber J, Rudobeck E, Campbell-Beachler M, Allen AR, Allen B, Rosi S, Nelson GA, Ramachandran S, Turner J, Fike JR, Vlkolinsky R (2014) (28)silicon radiation-induced enhancement of synaptic plasticity in the hippocampus of naive and cognitively tested mice. Radiat Res 181(4):362–368

Raber J, Allen AR, Weber S, Chakraborti A, Sharma S, Fike JR (2016) Effect of behavioral testing on spine density of basal dendrites in the CA1 region of the hippocampus modulated by (56)Fe irradiation. Behav Brain Res 302:263–268

Radford IR, Murphy TK, Radley JM, Ellis SL (1994) Radiation response of mouse lymphoid and myeloid cell lines. Part II. Apoptotic death is shown by all lines examined. Int J Radiat Biol 65(2):217–227

Ramadan SS, Sridharan V, Koturbash I, Miousse IR, Hauer-Jensen M, Nelson GA, Boerma M (2016) A priming dose of protons alters the early cardiac cellular and molecular response to (56)Fe irradiation. Life Sci Space Res 8:8–13

Rastegar N, Eckart P, Mertz M (2002) Radiation-induced cataract in astronauts and cosmonauts. Graefes Arch Clin Exp Ophthalmol 240(7):543–547

Reitz G, Hellweg CE (2018) Chapter 14: space radiation and its biological effects. In: Bolton PR, Parodi K, Schreiber J (eds) Applications of laser-driven particle acceleration. Taylor and Francis, an imprint of CRC Press, Boca Raton, FL, pp 217–237

Reynolds RJ, Day SM (2017) Mortality due to cardiovascular disease among apollo lunar astronauts. Aerosp Med Hum Perform 88(5):492–496

Rivera PD, Shih HY, Leblanc JA, Cole MG, Amaral WZ, Mukherjee S, Zhang S, Lucero MJ, Decarolis NA, Chen BP, Eisch AJ (2013) Acute and fractionated exposure to high-LET (56)Fe HZE-particle radiation both result in similar long-term deficits in adult hippocampal neurogenesis. Radiat Res 180(6):658–667

Rodemann HP, Peterson HP, Schwenke K, von Wangenheim KH (1991) Terminal differentiation of human fibroblasts is induced by radiation. Scanning Microsc 5(4):1135–1142. discussion 1142-1133

Rodier F, Campisi J (2011) Four faces of cellular senescence. J Cell Biol 192(4):547–556

Rogakou EP, Pilch DR, Orr AH, Ivanova VS, Bonner WM (1998) DNA double-stranded breaks induce histone H2AX phosphorylation on serine 139. J Biol Chem 273(10):5858–5868

Rogakou EP, Boon C, Redon C, Bonner WM (1999) Megabase chromatin domains involved in DNA double-strand breaks in vivo. J Cell Biol 146(5):905–916

Rola R, Sarkissian V, Obenaus A, Nelson GA, Otsuka S, Limoli CL, Fike JR (2005) High-LET radiation induces inflammation and persistent changes in markers of hippocampal neurogenesis. Radiat Res 164(4 Pt 2):556–560

Rothkamm K, Löbrich M (2003) Evidence for a lack of DNA double-strand break repair in human cells exposed to very low x-ray doses. Proc Natl Acad Sci USA 100(9):5057–5062

Rydberg B (1996) Clusters of DNA damage induced by ionizing radiation: formation of short DNA fragments. II. Experimental detection. Radiat Res 145(2):200–209

Sabatier L, Dutrillaux B, Martin MB (1992) Chromosomal instability. Nature 357(6379):548

Schöllnberger H, Stewart RD, Mitchel RE, Hofmann W (2004) An examination of radiation hormesis mechanisms using a multistage carcinogenesis model. Nonlinearity Biol Toxicol Med 2(4):317–352

Scholz M, Kraft-Weyrather W, Ritter S, Kraft G (1994) Cell cycle delays induced by heavy ion irradiation of synchronous mammalian cells. Int J Radiat Biol 66(1):59–75

Shikazono N, O'Neill P (2009) Biological consequences of potential repair intermediates of clustered base damage site in Escherichia coli. Mutat Res 669(1–2):162–168

Shiloh Y (2003) ATM and related protein kinases: safeguarding genome integrity. Nat Rev Cancer 3(3):155–168

Shimizu Y, Kodama K, Nishi N, Kasagi F, Suyama A, Soda M, Grant EJ, Sugiyama H, Sakata R, Moriwaki H, Hayashi M, Konda M, Shore RE (2010) Radiation exposure and circulatory disease risk: Hiroshima and Nagasaki atomic bomb survivor data, 1950–2003. BMJ 340:b5349

Shukitt-Hale B, Casadesus G, McEwen JJ, Rabin BM, Joseph JA (2000) Spatial learning and memory deficits induced by exposure to iron-56-particle radiation. Radiat Res 154(1):28–33

Shukitt-Hale B, Casadesus G, Cantuti-Castelvetri I, Rabin BM, Joseph JA (2003) Cognitive deficits induced by 56Fe radiation exposure. Adv Space Res 31(1):119–126

Shukitt-Hale B, Carey AN, Jenkins D, Rabin BM, Joseph JA (2007) Beneficial effects of fruit extracts on neuronal function and behavior in a rodent model of accelerated aging. Neurobiol Aging 28(8):1187–1194

Sieber OM, Heinimann K, Tomlinson IP (2003) Genomic instability—the engine of tumorigenesis? Nat Rev Cancer 3(9):701–708

Silasi G, Diaz-Heijtz R, Besplug J, Rodriguez-Juarez R, Titov V, Kolb B, Kovalchuk O (2004) Selective brain responses to acute and chronic low-dose X-ray irradiation in males and females. Biochem Biophys Res Commun 325(4):1223–1235

Singer M, De Santis V, Vitale D, Jeffcoate W (2004) Multiorgan failure is an adaptive, endocrine-mediated, metabolic response to overwhelming systemic inflammation. Lancet 364(9433):545–548

Smits R, Kartheuser A, Jagmohan-Changur S, Leblanc V, Breukel C, de Vries A, van Kranen H, van Krieken JH, Williamson S, Edelmann W, Kucherlapati R, KhanPm, Fodde R (1997) Loss of Apc and the entire chromosome 18 but absence of mutations at the Ras and Tp53 genes in intestinal tumors from Apc1638N, a mouse model for Apc-driven carcinogenesis. Carcinogenesis 18(2):321–327

Snyder AR, Morgan WF (2004) Gene expression profiling after irradiation: clues to understanding acute and persistent responses? Cancer Metastasis Rev 23(3–4):259–268

Sokolov MV, Dickey JS, Bonner WM, Sedelnikova OA (2007) Gamma-H2AX in bystander cells: not just a radiation-triggered event, a cellular response to stress mediated by intercellular communication. Cell Cycle 6(18):2210–2212

Soucy KG, Lim HK, Kim JH, Oh Y, Attarzadeh DO, Sevinc B, Kuo MM, Shoukas AA, Vazquez ME, Berkowitz DE (2011) HZE (5)(6)Fe-ion irradiation induces endothelial dysfunction in rat aorta: role of xanthine oxidase. Radiat Res 176(4):474–485

Sridharan DM, Asaithamby A, Blattnig SR, Costes SV, Doetsch PW, Dynan WS, Hahnfeldt P, Hlatky L, Kidane Y, Kronenberg A, Naidu MD, Peterson LE, Plante I, Ponomarev AL, Saha J, Snijders AM, Srinivasan K, Tang J, Werner E, Pluth JM (2016) Evaluating biomarkers to model cancer risk post cosmic ray exposure. Life Sci Space Res (Amst) 9:19–47

Stenerlow B, Hoglund E, Carlsson J, Blomquist E (2000) Rejoining of DNA fragments produced by radiations of different linear energy transfer. Int J Radiat Biol 76(4):549–557

Stoll U, Schmidt A, Schneider E, Kiefer J (1995) Killing and mutation of Chinese hamster V79 cells exposed to accelerated oxygen and neon ions. Radiat Res 142(3):288–294

Stoll U, Barth B, Scheerer N, Schneider E, Kiefer J (1996) HPRT mutations in V79 Chinese hamster cells induced by accelerated Ni, au and Pb ions. Int J Radiat Biol 70(1):15–22

Suzuki M, Watanabe M, Suzuki K, Nakano K, Kaneko I (1989) Neoplastic cell transformation by heavy ions. Radiat Res 120(3):468–476

Suzuki M, Tsuruoka C, Kanai T, Kato T, Yatagai F, Watanabe M (2003) Qualitative and quantitative difference in mutation induction between carbon- and neon-ion beams in normal human cells. Biol Sci Space 17(4):302–306

Sweet TB, Hurley SD, Wu MD, Olschowka JA, Williams JP, O'Banion MK (2016) Neurogenic effects of low-dose whole-body HZE (Fe) ion and gamma irradiation. Radiat Res 186(6):614–623

Sylvester CB, Abe JI, Patel ZS, Grande-Allen KJ (2018) Radiation-induced cardiovascular disease: mechanisms and importance of linear energy transfer. Front Cardiovasc Med 5:5

Tang FR, Loke WK (2015) Molecular mechanisms of low dose ionizing radiation-induced hormesis, adaptive responses, radioresistance, bystander effects, and genomic instability. Int J Radiat Biol 91(1):13–27

Tapio S, Jacob V (2007) Radioadaptive response revisited. Radiat Environ Biophys 46(1):1–12

Tenkorang MA, Snyder B, Cunningham RL (2018) Sex-related differences in oxidative stress and neurodegeneration. Steroids 133:21–27

Thacker J, Stretch A, Stephens MA (1979) Mutation and inactivation of cultured mammalian cells exposed to beams of accelerated heavy ions. II. Chinese hamster V79 cells. Int J Radiat Biol Relat Stud Phys Chem Med 36(2):137–148

Tofilon PJ, Fike JR (2000) The radioresponse of the central nervous system: a dynamic process. Radiat Res 153(4):357–370

Tornaletti S (2009) DNA repair in mammalian cells: transcription-coupled DNA repair: directing your effort where it's most needed. Cell Mol Life Sci 66(6):1010–1020

Tseng BP, Lan ML, Tran KK, Acharya MM, Giedzinski E, Limoli CL (2013) Characterizing low dose and dose rate effects in rodent and human neural stem cells exposed to proton and gamma irradiation. Redox Biol 1(1):153–162

Tseng BP, Giedzinski E, Izadi A, Suarez T, Lan ML, Tran KK, Acharya MM, Nelson GA, Raber J, Parihar VK, Limoli CL (2014) Functional consequences of radiation-induced oxidative stress in cultured neural stem cells and the brain exposed to charged particle irradiation. Antioxid Redox Signal 20(9):1410–1422

van der Burg M, van Dongen JJ, van Gent DC (2009) DNA-PKcs deficiency in human: long predicted, finally found. Curr Opin Allergy Clin Immunol 9(6):503–509

Van der Meeren A, Monti P, Vandamme M, Squiban C, Wysocki J, Griffiths N (2005) Abdominal radiation exposure elicits inflammatory responses and abscopal effects in the lungs of mice. Radiat Res 163(2):144–152

Villasana L, Acevedo S, Poage C, Raber J (2006) Sex- and APOE isoform-dependent effects of radiation on cognitive function. Radiat Res 166(6):883–891

Villasana L, Rosenberg J, Raber J (2010) Sex-dependent effects of 56Fe irradiation on contextual fear conditioning in C57BL/6J mice. Hippocampus 20(1):19–23

Villasana LE, Benice TS, Raber J (2011) Long-term effects of 56Fe irradiation on spatial memory of mice: role of sex and apolipoprotein E isoform. Int J Radiat Oncol Biol Phys 80(2):567–573

Villasana LE, Rosenthal RA, Doctrow SR, Pfankuch T, Zuloaga DG, Garfinkel AM, Raber J (2013) Effects of alpha-lipoic acid on associative and spatial memory of sham-irradiated and 56Fe-irradiated C57BL/6J male mice. Pharmacol Biochem Behav 103(3):487–493

Villasana LE, Weber S, Akinyeke T, Raber J (2016) Genotype differences in anxiety and fear learning and memory of WT and ApoE4 mice associated with enhanced generation of hippocampal reactive oxygen species. J Neurochem 138(6):896–908

Vlkolinsky R, Titova E, Krucker T, Chi BB, Staufenbiel M, Nelson GA, Obenaus A (2010) Exposure to 56Fe-particle radiation accelerates electrophysiological alterations in the hippocampus of APP23 transgenic mice. Radiat Res 173(3):342–352

Wang JYJ, Cho SK (2004) Coordination of repair, checkpoint, and cell death responses to DNA damage. Adv Protein Chem 69:101–135

Ward JF (1994) The complexity of DNA damage: relevance to biological consequences. Int J Radiat Biol 66(5):427–432

Werner E, Wang H, Doetsch PW (2014) Opposite roles for p38MAPK-driven responses and reactive oxygen species in the persistence and resolution of radiation-induced genomic instability. PLoS One 9(10):e108234

Whalen MK, Gurai SK, Zahed-Kargaran H, Pluth JM (2008) Specific ATM-mediated phosphorylation dependent on radiation quality. Radiat Res 170(3):353–364

White CR, Yu T, Gupta K, Babitz SK, Black LL, Kabarowski JH, Kucik DF (2015) Early changes in vascular reactivity in response to 56Fe irradiation in ApoE−/− mice. Acta Astronaut 108:40–45

Whoolery CW, Walker AK, Richardson DR, Lucero MJ, Reynolds RP, Beddow DH, Clark KL, Shih HY, LeBlanc JA, Cole MG, Amaral WZ, Mukherjee S, Zhang S, Ahn F, Bulin SE, DeCarolis NA, Rivera PD, Chen BPC, Yun S, Eisch AJ (2017) Whole-body exposure to (28) Si-radiation dose-dependently disrupts dentate gyrus neurogenesis and proliferation in the short term and new neuron survival and contextual fear conditioning in the long term. Radiat Res 188(5):532–551

Williams R (2015) NASA space flight human-system standard volume 1, revision A: crew health. NASA Technical Standards. National Aeronautics and Space Administration, Washington, DC. 1: 83

Wilson GD (2004) Radiation and the cell cycle, revisited. Cancer Metastasis Rev 23(3-4):209–225

Worgul BV, Merriam GR Jr, Medvedovsky C, Brenner DJ (1989) Accelerated heavy particles and the lens. III. Cataract enhancement by dose fractionation. Radiat Res 118(1):93–100

Worgul BV, Kundiev Y, Chumak VV, Ruban A, Parkhomenko G, Vitte P, Sergienko NM, Shore R, Likhtarev IA, Medvedovsky C, Junk AK (1999) The Ukranian/American Chernobyl ocular study. Ocular radiation risk assessment in populations exposed to environmental radiation contamination. A. K. Junk, Y. Kundiev, P. Vitte and B. V. Worgul. Dordrecht / Boston / London, Kluwer academic publishers. 2. Environment 50:1–12

Worgul BV, Brenner DJ, Medvedovsky C, Merriam GR Jr, Huang Y (1993) Accelerated heavy particles and the lens. VII: the cataractogenic potential of 450 MeV/amu iron ions. Invest Ophthalmol Vis Sci 34(1):184–193

Wu ZH, Miyamoto S (2007) Many faces of NF-kappaB signaling induced by genotoxic stress. J Mol Med (Berl) 85(11):1187–1202

Wu B, Medvedovsky C, Worgul BV (1994) Non-subjective cataract analysis and its application in space radiation risk assessment. Adv Space Res 14(10):493–500

Wu H, Durante M, George K, Yang TC (1997) Induction of chromosome aberrations in human cells by charged particles. Radiat Res 148(5 Suppl):S102–S107

Wyman C, Kanaar R (2006) DNA double-strand break repair: all's well that ends well. Annu Rev Genet 40:363–383

Wyrobek AJ, Britten RA (2016) Individual variations in dose response for spatial memory learning among outbred wistar rats exposed from 5 to 20 cGy of (56) Fe particles. Environ Mol Mutagen 57(5):331–340

Yang VC, Ainsworth EJ (1987) A histological study on the cataractogenic effects of heavy charged particles. Proc Natl Sci Counc Repub China B 11(1):18–28

Yatagai F (2004) Mutations induced by heavy charged particles. Biol Sci Space 18(4):224–234

Yeiser LA, Villasana LE, Raber J (2013) ApoE isoform modulates effects of cranial (5)(6)Fe irradiation on spatial learning and memory in the water maze. Behav Brain Res 237:207–214

Yu T, Parks BW, Yu S, Srivastava R, Gupta K, Wu X, Khaled S, Chang PY, Kabarowski JH, Kucik DF (2011) Iron-ion radiation accelerates atherosclerosis in apolipoprotein E-deficient mice. Radiat Res 175(6):766–773

Chapter 4
Radiation Risk Assessment

Abstract The assessment of risks associated with space radiation exposure requires determination of the expected radiation exposure during the space mission, knowledge of the risks associated with such exposures and of modifying factors which are all to be incorporated in a dedicated radiation risk model. To determine the radiation exposure, the radiation field and its variation during the solar cycle, its interaction with matter and the dose distribution in the human body are considered. Disease risks are derived from large epidemiological studies on radiation-exposed populations, predominantly the Japanese atomic bomb survivors. Extrapolation from high dose rate to low-dose rate exposure, from low linear energy transfer (LET) to high-LET radiation and from the Japanese to other populations requires the use of extrapolation factors. Accordingly, predictions of cancer risk and acceptable radiation exposure in space are subject to many uncertainties including the relative biological effectiveness (RBE) of space radiation (especially heavy ions), dose rate effects and possible interaction with microgravity and other spaceflight environmental factors.

Keywords Galactic cosmic rays (GCR) · Solar particle events (SPE) · Biological weighting factors · NASA Space Cancer Risk (NSCR) model · Excess relative risk · Dose and dose-rate reduction effectiveness factor (DDREF) · Risk of Exposure-Induced Death (REID) · Acceptable risk level

4.1   Radiation Exposure

For assessment of the risk to develop a disease and even die from this disease after space radiation exposure, a complex evaluation of physical, biological, and epidemiological data and the development of a model based on various assumptions are necessary.

First of all, the dose that astronauts are exposed to has to be determined. Therefore, the composition of the radiation field and its interaction with the spacecraft or habitat hull has to be calculated. Galactic cosmic radiation (GCR) environmental models and "particle transport codes describing the GCR modification by atomic and nuclear interactions in spacecraft and tissue shielding" are used for calculation (National Research Council 2012; Cucinotta et al. 2013a).

In low Earth orbit (LEO), e.g., on the International Space Station (ISS), trapped protons contribute to the accumulated dose. The variation of the solar activity during the approximately 11-year solar cycle has to be considered as it strongly affects the GCR flux, with a factor of up to three between solar maximum and solar minimum when solar modulation of GCR is weakest, resulting in a variation of GCR organ exposures of about two-fold and thereby a two-fold lower risk at solar maximum compared to solar minimum (NCRP 2006; Cucinotta et al. 2013a). On the other hand, a response to occasional solar particle events (SPE) might be necessary. On planetary surfaces, the atmosphere can reduce the dose rates and buildup of particles has to be factored in. In the NASA model, the GCR descriptive parameter is Z^{*2}/β^2 (describes the density of the ionization of a particle track), where Z^* is a particle's effective charge number and β its velocity relative to the speed of light.

Concerning the shielding of the spaceship or the habitat, the material composition, density, and thickness influence how the radiation field is altered and how strong the secondary particle buildup is. For GCR, a near balance between the loss of energetic particles and the new production of ions through atomic and nuclear interaction can be expected for the surface of Mars (Cucinotta et al. 2013a).

Next, the interaction with the human body comes into play. Organ doses or tissue average absorbed doses resulting from travel of the space radiation's particles through the body have to be weighted with the corresponding tissue weighting and quality factor.

4.2   Disease Risks

In this step, the disease risks for the expected exposure are calculated. For different diseases such as cancer and cardiovascular diseases, disease risks for exposures to low-LET radiation were derived from epidemiological data of human exposures (atomic bomb explosions in Japan; occupational exposures). To extrapolate the disease risks from human exposures to low-LET radiation to high-LET radiation, different biological weighting factors are used (see Sect. 3.3). Furthermore, for the Japanese data, an extrapolation to the population of the astronauts (the USA,

European, or other) is required (Durante and Cucinotta 2008). Radiation risks that are proportional to spontaneous cancer risks are assumed in the multiplicative transfer model[1], while an independent action of radiation is presumed in the additive transfer model[2] (Cucinotta et al. 2005). The National Council on Radiation Protection and Measurements (NCRP) recommends a mixture of both, the multiplicative risk model and additive risk model for transferal of cancer risk estimates from the Japanese to the US population (NCRP 2000; Cucinotta et al. 2005).

The risk is modified by many factors, e.g., sex, age at the time of radiation exposure, the time that passed since the exposure, and the attained age. The extent of risk modification by these factors depends on the cancer type. Also, interactions with other risk factors for cancerogenesis can affect the radiation-related excess risk, such as smoking, being overweight, or excessive alcohol consumption. It is unclear whether or to what extent individual radiation sensitivity modifies the disease risk.

The NASA Space Cancer Risk (NSCR) model estimates cancer risks and uncertainties for human space missions (Cucinotta et al. 2013a). Fatal cancer risk is regarded as the dominant risk arising from GCR exposure (National Research Council 1996, 2012; Cucinotta et al. 2013a). In the NASA model, the dose equivalents for leukemia and solid cancer risks for HZE particles are calculated with the quality factor (Q, see Sects. 2.2.1 and 3.3) for cancer. A never-smoker population is used to represent astronauts (Cucinotta et al. 2013a). Q is based on various experiments determining the RBE for tumor induction by accelerated ions or different LET (ICRP 2003), including the RBE for induction of Harderian gland tumors by Fe particles in mice, which amounts to 30, and the RBE for liver tumors in mice, which exceeds 50 (Cucinotta et al. 2013a). These data were recently completed with low dose exposures to silicon and titanium ions (Chang et al. 2016). RBEs for cancer in other tissues are scarce (Table 3.1). NASA's space radiation quality factor (QF) is based on track structure concepts, considering the (particle kinetic energy per nucleon (E) and the charge number (Z) instead of the LET (National Research Council 2012; Cucinotta et al. 2013a, 2019; NCRP 2014; Cucinotta 2014, 2015). Attempts to include non-targeted effects with their nonlinear dose–response relationship (see Sect. 3.2) in modeling of tumor prevalence as effect of radiation exposure resulted in a predicted risk that is two-fold higher compared to a targeted effects model (Cucinotta and Cacao 2017).

Recently, a dynamical model of the assessment of the excess relative risk (ERR)[3] for radiogenic leukemia among acutely/continuously irradiated humans was applied for ERR estimation of astronauts on interplanetary missions, which considers also

---

[1] In the multiplicative transfer model, the effect of the covariates on cancer is to act multiplicatively on a baseline hazard rate.

[2] In the additive model, covariates act in an additive manner on a baseline hazard rate.

[3] The ERR is the proportional increase in risk over the absolute risk in the absence of exposure (background). The background cancer risks can differ between populations, e.g., Japanese and US population.

granulo- and lymphopoiesis and can be applied for emergency cases of high SPE exposures (Smirnova and Cucinotta 2018).

To determine the non-cancer risk dose equivalent, the RBE for circulatory disease is used, based on the deterministic effects RBE model by ICRP (ICRP 1990). Such factors are not yet defined for degenerative alterations of the CNS.

4.3   Extrapolation Factors

Depending on the data underlying factors such as Q, QF, or RBE (see Sect. 3.3)—meaning whether they are derived from acute or chronic exposure, a correction is necessary for extrapolation from acute high dose and high-dose rate exposures to exposures to low doses and at low dose rates. This is accomplished by means of a dose and dose-rate reduction effectiveness factor (DDREF). Here, the track structure with a low ionization density part in the track penumbra and a high ionization density section in the track core has to be considered (Sect. 3.2.1), whereby only for the low ionization density part, dose rate effects are expected (Cucinotta et al. 2019). Based on the BEIR VII report (National Research Council 2006) and recommendations of the National Research Council (NRC) to NASA (National Research Council 2012), a factor of 1.5 is used for cancer risk predictions (Cucinotta et al. 2013a).

Unfortunately, DDREF[4] for cardiovascular and CNS disease risk are not available for heavy ions at relevant doses and dose rates (NCRP 2006; Cucinotta et al. 2013a).

4.4   Risk Uncertainties and Acceptability

The various inputs leading to the risk of radiation exposure-induced death (REID) are summarized in Fig. 4.1. The value of Q or QF (in the NSCR model) contributes strongly to the uncertainty of REID (National Research Council 2012).

In spaceflight, possible interactions of radiation with other spaceflight environmental factors have to be considered as possibly additive or even synergistic factors in cancerogenesis. These factors encompass microgravity causing fluid shifts in the body and muscle and bone degradation, acceleration forces, vibration, noise, stress,

---

[4]The term dose and dose rate effectiveness factor (DDREF) has a significant impact on space radiation protection requirements, especially for operational implementation. It is used to modify the dose-risk relationship estimated from the LNT model for (in the current respect space-typical) exposure scenarios: (1) the dose and (2) dose rate at which the dose is delivered. Use of the DDREF thereby reduces the lifetime attributable risk (LAR) of cancer incidence. From operational view, the BEIR (Biological Effects of Ionizing Radiation) VII Committee set the DDREF at a value of 2. Calabrese, E. J. and M. K. O'Connor (2014). "Estimating risk of low radiation doses—a critical review of the BEIR VII report and its use of the linear no-threshold (LNT) hypothesis." *Radiat Res* **182**(5): 463–474.

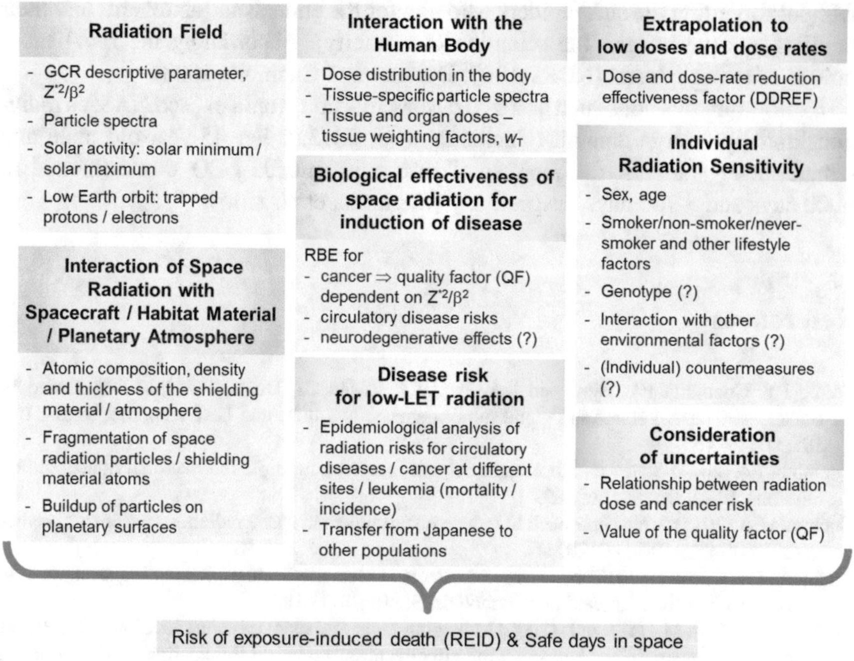

Radiation Field	Interaction with the Human Body	Extrapolation to low doses and dose rates
- GCR descriptive parameter, Z^{*2}/β^2 - Particle spectra - Solar activity: solar minimum / solar maximum - Low Earth orbit: trapped protons / electrons	- Dose distribution in the body - Tissue-specific particle spectra - Tissue and organ doses – tissue weighting factors, w_T	- Dose and dose-rate reduction effectiveness factor (DDREF)

	Biological effectiveness of space radiation for induction of disease	Individual Radiation Sensitivity
Interaction of Space Radiation with Spacecraft / Habitat Material / Planetary Atmosphere	RBE for - cancer ⇒ quality factor (QF) dependent on Z^{*2}/β^2 - circulatory disease risks - neurodegenerative effects (?)	- Sex, age - Smoker/non-smoker/never-smoker and other lifestyle factors - Genotype (?) - Interaction with other environmental factors (?) - (Individual) countermeasures (?)
- Atomic composition, density and thickness of the shielding material / atmosphere - Fragmentation of space radiation particles / shielding material atoms - Buildup of particles on planetary surfaces	**Disease risk for low-LET radiation** - Epidemiological analysis of radiation risks for circulatory diseases / cancer at different sites / leukemia (mortality / incidence) - Transfer from Japanese to other populations	**Consideration of uncertainties** - Relationship between radiation dose and cancer risk - Value of the quality factor (QF)

Risk of exposure-induced death (REID) & Safe days in space

Fig. 4.1 Risk assessment for space radiation exposure of astronauts during space missions in LEO and beyond. The graph is based on the NASA Space Cancer Risk (NSCR) model and its evaluation by an expert group (National Research Council 2012). According to Cucinotta et al. (2013b), "Demographic Specific Solar Minimum Safe Days in deep space are defined as the maximum number of days with 95% CL to be below the NASA 3% REID limit for males and females at different ages at exposure, aE." Z^* is the effective charge number of the particle and β its speed relative to the speed of light

and living in a closed environment (with, e.g., elevated carbon dioxide levels or in an exploration atmosphere with reduced pressure and a higher oxygen percentage) and with an externally determined time schedule leading to altered biorhythms.

According to the Aerospace Safety Advisory Panel, the human space missions should be designed to keep the risk for during mission loss of crew (LOC) below 1 in 270, and the long-term goal is to reach 1 in 750 (Dyer 2011). The radiation risks contribute to this aggregate risk of a space mission (Cucinotta et al. 2013a). NASA's limit for astronauts' space radiation exposure is set to an exposure-induced death risk (REID) of 3%, which has to be judged at the upper 95% confidence interval (CI) of the risk estimate (National Aeronautics and Space Administration (NASA) 2007; Durante and Cucinotta 2008; Cucinotta et al. 2013a).

The 3 % exposure-related fatality risk limit adopted by NASA is derived from data obtained in the 1980s for work-place related worker fatalities in less-safe industries (Cucinotta et al. 2019).

According to the NCRP, the acceptable risk level has to consider the value of the mission in terms of gain of knowledge and technology for humanity and science without forgetting the stakeholders who pay for the endeavour (astronauts and their families, tax-payers and the scientific community). Accordingly the risk-benefit ratio is of major interest (NCRP 1997; Durante and Cucinotta 2008).

Exploration missions such as a 940 d Mars mission would exceed NASA's radiation limit by a large amount (Cucinotta et al. 2013a). For 45-year-old male and female never smokers, the number of safe days outside LEO was estimated to >900 days and > 700 days, respectively (Cucinotta et al. 2015).

References

Chang PY, Cucinotta FA, Bjornstad KA, Bakke J, Rosen CJ, Du N, Fairchild DG, Cacao E, Blakely EA (2016) Harderian gland tumorigenesis: low-dose and LET response. Radiat Res 185(5):449–460

Cucinotta FA (2014) Space radiation risks for astronauts on multiple international Space Station missions. PLoS One 9(4):e96099

Cucinotta FA (2015) A new approach to reduce uncertainties in space radiation cancer risk predictions. PLoS One 10(3):e0120717

Cucinotta FA, Cacao E (2017) Non-targeted effects models predict significantly higher mars mission cancer risk than targeted effects models. Sci Rep 7(1):1832

Cucinotta FA, Kim M-HY, Ren L (2005) Managing lunar and mars mission radiation risks. Part 1; cancer risks, uncertainties, and shielding effectiveness. Technical Publication, NASA Johnson Space Center, Houston, TX, p 44

Cucinotta F, Kim M, Chappell L (2013a) Space radiation cancer risk projections and uncertainties-2012. NASA/TP-2013-217375. NASA, Houston, TX

Cucinotta FA, Kim MH, Chappell LJ, Huff JL (2013b) How safe is safe enough? Radiation risk for a human mission to Mars. PLoS One 8(10):e74988

Cucinotta FA, Alp M, Rowedder B, Kim MH (2015) Safe days in space with acceptable uncertainty from space radiation exposure. Life Sci Space Res (Amst) 5:31–38

Cucinotta FA, Cacao E, Kim MY, Saganti PB (2019) Non-targeted effects lead to a paridigm shift in risk assessment for a mission to the earth's moon or martian moon phobos. Radiat Prot Dosim 183(1–2):213–218

Durante M, Cucinotta FA (2008) Heavy ion carcinogenesis and human space exploration. Nat Rev Cancer 8(6):465–472

Dyer J (2011) Aerospace safety advisory panel annual report for 2010. Washington, DC

ICRP (1990) RBE for deterministic effects. ICRP Publication 58. Ann. ICRP 20 (4). F. D. Sowby, International Commission on Radiological Protection

ICRP (2003) Relative biological effectiveness (RBE), quality factor (Q), and radiation weighting factor (wR). ICRP Publication 92

National Aeronautics and Space Administration (NASA), W., DC (2007) NASA Space flight human system standard volume I; Crew health. NASA STD-3001. National Aeronautics and Space Administration (NASA), Washington, DC. NASA-STD-3001: 83

National Research Council (1996) Radiation hazards to crews of interplanetary missions: biological issues and research strategies. The National Academies Press, Washington, DC

National Research Council (2006) Health risks from exposure to low levels of ionizing radiation: BEIR VII phase 2. National Academies Press, Washington, DC

National Research Council (2012) Technical evaluation of the NASA model for cancer risk to astronauts due to space radiation. The National Academies Press, Washington, DC

NCRP (1997) National council on radiation protection and measurements: acceptability of risk from radiation: application to human space flight. NCRP, Bethesda, MD

NCRP (2000). NCRP Report No. 132: Radiation protection guidance for activities in low earth orbit. Bethesda, MD, USA

NCRP (2006) NCRP report no. 153: information needed to make radiation protection recommendations for space missions beyond low-earth orbit, vol 153. NCRP, Bethesda, MD, pp 1–427

NCRP (2014) Commentary no. 23—radiation protection for space activities: supplement to previous recommendations. NCRP, Bethesda, MD. Commentary No. 23: 1–427

Smirnova OA, Cucinotta FA (2018) Dynamical modeling approach to risk assessment for radiogenic leukemia among astronauts engaged in interplanetary space missions. Life Sci Space Res (Amst) 16:76–83

Chapter 5
Space Radiation Countermeasures

Abstract Different countermeasures against deleterious effects of space radiation on human health are required for galactic cosmic rays (GCR) and solar particle events (SPE). In case of SPE, efficient shielding is possible in a radiation shelter. The crew has to enter the shelter in case of increasing proton fluxes during an SPE to avoid high-dose exposures within hours or days which might lead to the acute radiation syndrome (ARS) in worst case scenarios. ARS treatment strategies are available but difficult to implement on a space mission. The chronic exposure to GCR is associated with various health risks (cancer, cataract, CNS decrements, cardiovascular disease) that can reduce the healthy life span after return to Earth. Shielding of GCR in free space is very limited and therefore, the time in free space should be minimized by fast propulsion. On planetary and moon surfaces, the habitats should provide optimal shielding in order to reduce GCR dose accumulated during the mission. Much effort is placed on the development of nutritional and pharmaceutical countermeasures for chronic and late radiation effects. Free radical scavengers such as amifostine are ruled out because of severe side effects. Encouraging results were reported in rodent studies with berries or dried plums. A concerted action of ground-based studies and space experiments is required to expand radiobiological knowledge about space radiation to improve countermeasure development. Crew selection is a sensitive topic, but from a radiation protection point of view, the lower radiosensitivity of older crew members might be considered.

Keywords Radiation protection principles · Shielding · Hibernation · Acute Radiation Syndrome (ARS) treatment · Amifostine · Antioxidants · Phytochemicals · Ataxia telangiectasia mutated (ATM) gene

5.1 Introduction

The usual recommendations for radiation protection on Earth are to reduce the exposure time, to increase the distance to the radiation source, to use shielding, and to reduce the activity of the radiation source. The activity of the Sun and the intensity of galactic cosmic radiation (GCR) are beyond the human sphere of influence. As space radiation in free space impacts from all directions, increasing the distance plays no role in space radiation protection (Durante and Cucinotta 2008). Exposure time in free space during a Mars mission could be reduced by faster propulsion of the space vehicle, which is not expected to be possible within a realistic time period. In habitats on other planets or moons, a dose reduction could be achieved by shielding using materials available on site. For the spaceship, GCR shielding is impossible or at least very limited because of current launch mass capabilities (Durante and Cucinotta 2008). Therefore, much hope is set in the development of pharmacological or nutritional countermeasures for space radiation effects, but a major breakthrough in cancer prevention is currently not unveiling. So, as a last way out, selection of radio-resistant individuals for exploration missions and use of hibernation are discussed. Independent of the question whether humans can achieve a hibernation state, it is very questionable whether a state of reduced metabolic activity is beneficial in case of a chronic low-dose exposure, as protective effects were shown only for acute effects of high-dose exposures (Cerri et al. 2016; Tinganelli et al. 2019).

5.2 Operational Planning and Shielding

The best radiation protection, not only in space, is the avoidance or reduction of exposure through careful mission planning and shielding. The shielding of a spaceship is in the range of 5 to 20 cm of water equivalent (5 to 20 g/cm^2), while the shielding of the space suit of an astronaut shrinks on average to 1 cm of water equivalent (1 g/cm^2) during a spacewalk.

Shielding is highly effective against the solar component of cosmic radiation, and especially against its low-energy components. The Orion spaceship of the NASA Artemis program (formerly Exploration Mission, EM) disposes of a radiation shelter which the crew will enter in case of a SPE warning.

The reduction of the GCR dose that can be achieved with realistic thicknesses of aluminum shielding is limited for the high-energy particles of the galactic component. After a rapid decrease of the dose equivalent caused by fragmentation of the primary GCR and a corresponding decrease in the quality factor in the initial 10–20 g/cm^2, the dose equivalent rate is expected to change only slowly with further increasing of shielding. In model calculations with water and aluminum shields, the risk of cancer death decreased barely with increasing shielding thickness up to 150 g/cm^2 (Durante and Cucinotta 2008).

Lightweight high hydrogen content materials, e.g., polyethylene, are much more effective for shielding; nevertheless, since the weight of the shield cannot be increased indefinitely, it is expected that the reduction in the dose that can be reached using mass shielding for a spacecraft is no more than 50%. In addition to suitable radiation protection rooms, personal radiation protection vests are also tested.

5.3   Prevention and Therapy of the Acute Radiation Syndrome (ARS)

Therapeutic approaches for the ARS were tested in non-human primates, dogs, mice, rats, and pigs (Donnadieu-Claraz et al. 1999; MacVittie et al. 2005) and in accidently irradiated humans. Based on clinical findings, they include isolation of the patient to reduce the danger of infection, antibiotics in combination with an antifungal drug if an infection occurs, electrolyte and platelet or blood transfusions, and allogenic bone marrow transplantation from donors with compatible surface antigens (Cronkite 1964). Current therapeutical standards include the stimulation of remaining bone marrow stem cells by Granulocyte Colony-Stimulating Factor (G-CSF) after exposure to less than 50% of the lethal dose (Romero-Weaver et al. 2014), and allogenic stem cell transplantation if the bone marrow is completely eradicated (Drouet and Herodin 2010). The efficacy of other growth factors, such as Keratinocyte Growth Factor (KGF) and combinations of different growth factors (Stem Cell Factor—SCF, Nerve Growth Factor—NGF, erythropoietin, PEGylated growth factors), antioxidants (e.g., N-acetyl cysteine), and anti-inflammatory approaches (e.g., inhibitors of cyclooxygenase-2—COX-2-, anti-IL-antibodies, curcumin, Ghrelin) and hypoxia are currently under investigation (Neal et al. 2003; Drouet and Herodin 2010; Jacob et al. 2010; Kiang and Olabisi 2019).

The activation of survival promoting intracellular signaling pathways (Burdelya et al. 2012; Toshkov et al. 2017) was suggested as ARS treatment strategy. Due to its prosurvival effects, the Nuclear Factor κB (NF-κB, Sect. 3.2) pathway was chosen as target; bacterial flagellin, the natural agonist of the NF-κB-activating Toll-like receptor 5 (TLR5), was effective in protecting tissues expressing this receptor from radiation damage (Gudkov and Komarova 2010). TLR5 is expressed by cells of the hematopoietic system and of the gastrointestinal tract tissues, which determine survival and severity of ARS. The effect of this flagellin-based drug after proton exposure (to simulate an SPE) remains to be determined.

The inhibition of apoptosis-promoting pathways such as the p53 pathway by the small-molecule inhibitor pifithrin α for the treatment of ARS might reduce massive radiation-induced cell death in the hematopoietic system such as spleen, thymus, bone marrow, and lymph nodes (Komarov et al. 1999). Unfortunately, p53 inhibitors were found to increase the radiosensitivity of the gastrointestinal tract (Burdelya et al. 2006).

After recovery from ARS, the probability of developing leukemia is about 100 cases per 10^5 population at risk (mostly closely exposed Nagasaki atomic bomb survivors) in the first 10–15 years after exposure (Tomonaga 1962).

ARS is a severe disease with a bad to infaust prognosis at higher doses even with intensive care. During an SPE, doses eliciting the hematological syndrome could be accumulated. This would require intense therapeutic care which is unrealistic during a space mission. Therefore, such exposures have to be avoided by early warning systems for SPE and appropriate radiation shelters. To provide sufficient redundancy in case of failure of the warning system, an ARS treatment kit could be carried anyway on a deep space mission.

5.4 Nutritional and Pharmaceutical Countermeasures for Chronic and Late Radiation Effects

As cancer risk by GCR is a major limiting factor in spaceflight, reduction of this risk by pharmacological or nutraceutical intervention is highly desirable. In a recent roadmap, NASA assigned a Technology Readiness Level (TRL) of 1 to countermeasures against degenerative effects, with the goal of TRL 9, underlining the urgency to develop and test biological treatments for chronic low-dose exposure to GCR (NASA 2015).

A drug or dietary intervention can be understood as a radiation protection measure. Organic thiophosphates such as amifostine (WR-2721) can act as free radical scavengers and protect tissues against harmful effects of ionizing radiation; it is therefore approved for clinical use during cancer treatments (Sasse et al. 2006; Kouvaris et al. 2007). The side effects of amifostine encompass nausea, vomiting, vasodilatation, and hypotension (Boccia 2002), and they limit the routine clinical use (Kamran et al. 2016). Therefore, a long-term use during space missions is excluded. On the one hand, a short-term administration in case of a life-threatening SPE is suggested to be beneficial (Durante and Cucinotta 2008), on the other hand, doubts remain concerning its safety and efficacy for prevention of ARS (Singh and Seed 2019).

In the search for antioxidants with lower toxicity which can be applied for longer time periods, nutritional antioxidants are the favorites regardless of a lack of effect of dietary antioxidant supplementation in several human diseases (Halliwell 2000, 2013; Bingham and Riboli 2004). Nevertheless, the European Prospective Investigation into Cancer and Nutrition (EPIC) showed inverse associations of fruit intake with upper gastrointestinal tract and lung cancer risk, of intakes of total fruits and vegetables and total fiber with colorectal cancer risk, and of intake of total fiber with liver cancer risk (Bradbury et al. 2014). On the other hand, a meta-analysis of 68 trials came to the conclusion that supplementation of beta carotene, vitamin A and E may increase mortality (Bjelakovic et al. 2007). Concerning the effect in a radiation-exposed cohort, a cancer risk reduction was found for the atomic bomb survivors with higher daily fruit and vegetable intake (Sauvaget et al. 2004). Furthermore, in some astronauts, the antioxidant capacity was reduced after an ISS mission (Smith et al. 2005; Durante and Cucinotta 2008).

Thus, the enrichment of food with radical scavengers and antioxidants for protection against reactive oxygen species (ROS, Sect. 3.2) is considered. In animal experiments, feeds fortified with berries (strawberries, blueberries) or dried plums have shown good results (Guan et al. 2004; Rabin et al. 2005a; Schreurs et al. 2016; Poulose et al. 2017), and antioxidants were also effective when they were administered after exposure with high-LET radiation (Kennedy 2014; Sridharan et al. 2016). The berry diet was effective in rodent models for deleterious CNS effects of heavy ions (Rabin et al. 2005b; Shukitt-Hale et al. 2007) and also reduced heavy ion-induced tumorigenesis (Rabin et al. 2005a). Other candidates are green tea and cruciferous plants, hormones (e.g., melatonin), glutathione, superoxide dismutase (SOD), metals (especially selenium, zinc, and copper salts) (Weiss and Landauer 2000; Durante and Cucinotta 2008) and inhibitors of the NF-κB pathway (nonsteroidal anti-inflammatory drugs (NSAIDs) (Yamamoto and Gaynor 2001), curcumin (Inano et al. 2000), black raspberries (Huang et al. 2007; Madhusoodhanan et al. 2010), and extracts of unripe kiwifruits (Fig. 5.1) and of apples (Davis et al. 2006; Abe et al. 2010). However, possible side effects of long-term inhibition of the NF-κB pathway need to be considered. The chronic and systemic use of strong "broadband" NF-κB inhibitors is well able to suppress immune activation in normal immune cells. This can lead to severe immunodeficiency, which is not tolerable in cancer prevention and human spaceflight (Wong and Tergaonkar 2009). Especially in astronauts, immune suppression in addition to the effects of microgravity on the immune system is not desirable.

What is valid on Earth in terms of chemoprevention of cancer holds also true for space missions: the risk might be reduced by a diet rich in phytochemicals (Kresty et al. 2001; Genkinger et al. 2004), but an absolute protection is not possible. Much more studies are necessary to find the optimal combination, pharmaceutical form, and dosage of the phytochemicals.

Fig. 5.1 Products of the secondary plant metabolism as nutritional countermeasure candidates. Kiwi, pomegranate, or other fruits and vegetables contain secondary metabolites which might have health-beneficial effects. (© Christine E. Hellweg, DLR)

5.5   Crew Selection and Personalized Prevention

From a bio-medical point of view, a sensitive selection of astronauts can lead to a reduction in health impairments caused by space radiation and its consequences. For older crew members, the expected loss of a healthy lifetime is lower than it is for younger people. In addition, astronauts beyond the reproductive age cannot contribute to an increase in the number of teratogenic or genetic damage in a population, as they no longer have any offspring to expect. For the induction of certain types of tumors, there is also a sex-specific relationship. Women might be at greater risk for radiation-induced cancer than men (National Research Council 2006; Narendran et al. 2019). This could favor the selection of male crews as women have an additional risk of breast cancer. On the other hand, from a psychological point of view, there are the advantages of a mixed crew. Furthermore, recent studies on neurodegenerative effects induced by heavy ion exposure have shown that female mice are less affected compared to male mice because of a different microglia reaction (Krukowski et al. 2018).

Last but not least, the existence of inter-individual variations in radiation sensitivity, as demonstrated for UV light by the various skin types, is far from clear for exposure to ionizing radiation. Here, confusion on the notion "radiosensitivity" exists (Britel et al. 2018). This term was coined in the beginnings of radiotherapy to describe skin reactions and later on other adverse tissue reactions which are considered to rely predominantly on cell death. To distinguish the increased sensitivity towards ionizing radiation-induced tissue reactions from a cancer-proneness in response to ionizing radiation exposure which is attributable to cell transformation, the term "radiosusceptibility" is suggested (Britel et al. 2018). This differentiation is useful for the clinical situations as radiosensitivity and radiosusceptibility not necessarily come along with each other: a patient who has a higher risk of a secondary cancer can have a normal tissue reaction in radiotherapy and vice versa. The genetic background is crucial for the characteristics of the sensitivity. The most prominent example for increased sensitivity to ionizing radiation are mutations in the *Ataxia telangiectasia* mutated (ATM) gene. ATM homozygotes are extremely radiosensitive individuals (Taylor et al. 1975), and they are also radiosusceptible. 0.5%–1% of the general population carry a heterozygote ATM mutation and are also cancer-prone and radiosensitive (Broeks et al. 2000; Thompson et al. 2005). Experiments with ATM homozygotes, heterozygotes, and wild-type mice revealed that the heterozygotes were more sensitive to X-ray-induced cataracts compared to wild-type mice (Worgul et al. 2002; Durante and Cucinotta 2008).

In contrast, patients with the Li Fraumeni syndrome ($p53^{+/-}$) are cancer-prone as the "guardian of the genome," p53, is impaired, but they are not hypersensitive to radiation in terms of tissue reactions.

10–15% of patients receiving radiotherapy suffer adverse reactions (Averbeck et al. 2020). Based on studies with radiotherapy patients, three groups of radiosensitivity were defined. In group III, the extreme radiosensitivity is caused by homozygous mutations of radiosensitivity genes which are directly involved in DNA

DSB repair or more generally, in the DNA damage response (Sect. 3.2.3, for example, Rad9 (Kleiman et al. 2007), BRCA1, and BRCA2 (Baeyens et al. 2004; Durante and Cucinotta 2008)). In group II, a moderate radiosensitivity is explained by mutations of these genes in heterozygous state (Copernic Project Investigators et al. 2016) and 2–15% of the patients display such moderate overreaction (Averbeck et al. 2020). Group I individuals display normal radiosensitivity (~ 85% of patients).

Group III individuals are affected by complex genetic syndromes with a spectrum of other defects, while group II individuals might not be affected by any visible defect or disease. Therefore, individuals with a reduced DNA DSB repair capability ("group II") might be present in astronaut corps. Here, it is important to know whether they are radiosusceptible—meaning that they have a high risk to develop radiation-induced tumors—and whether they are more susceptible to degenerative effects induced by ionizing radiation exposure ("radiodegeneration") (Foray et al. 2016). This knowledge will enable flight surgeons to develop preventive measures such as close meshed medical examinations.

For cancer patients, a large spectrum of tests exists to determine their radiosensitivity using a skin biopsy or blood sample (e.g., clonogenic cell survival tests with fibroblasts, DNA repair assays based on γHAX/pATM/MRE11/53BP1 immunofluorescence, micronuclei) (Ferlazzo et al. 2017). As these tests focus on tissue reactions as side effect of radiotherapy, there relevance for predicting increased sensitivity towards space radiation-induced late effects such as cancer, cataract, CNS alterations, and cardiovascular disease has to be evaluated before they can be used to offer astronauts a tool to determine their individual risk and based on that, personalized preventive measures.

References

Abe D, Saito T, Kubo Y, Nakamura Y, Sekiya K (2010) A fraction of unripe kiwi fruit extract regulates adipocyte differentiation and function in 3T3-L1 cells. Biofactors 36(1):52–59

National Research Council (2006) Health risks from exposure to low levels of ionizing radiation: BEIR VII phase 2. National Academies Press, Washington, DC

Averbeck D, Candéias S, Chandna S, Foray N, Friedl AA, Haghdoost S, Jeggo PA, Lumniczky K, Paris F, Quintens R, Sabatier L (2020) Establishing mechanisms affecting the individual response to ionizing radiation. Int J Radiat Biol 96(3):297–323

Baeyens A, Thierens H, Claes K, Poppe B, de Ridder L, Vral A (2004) Chromosomal radiosensitivity in BRCA1 and BRCA2 mutation carriers. Int J Radiat Biol 80(10):745–756

Bingham S, Riboli E (2004) Diet and cancer—the European prospective investigation into Cancer and nutrition. Nat Rev Cancer 4(3):206–215

Bjelakovic G, Nikolova D, Gluud LL, Simonetti RG, Gluud C (2007) Mortality in randomized trials of antioxidant supplements for primary and secondary prevention: systematic review and meta-analysis. JAMA 297(8):842–857

Boccia R (2002) Improved tolerability of amifostine with rapid infusion and optimal patient preparation. Semin Oncol 29(6 Suppl 19):9–13

Bradbury KE, Appleby PN, Key TJ (2014) Fruit, vegetable, and fiber intake in relation to cancer risk: findings from the European prospective investigation into Cancer and nutrition (EPIC). Am J Clin Nutr 100(Suppl 1):394s–398s

Britel M, Bourguignon M, Foray N (2018) The use of the term 'radiosensitivity' through history of radiation: from clarity to confusion. Int J Radiat Biol 94(5):503–512

Broeks A, Urbanus JH, Floore AN, Dahler EC, Klijn JG, Rutgers EJ, Devilee P, Russell NS, van Leeuwen FE, van't Veer LJ (2000) ATM-heterozygous germline mutations contribute to breast cancer-susceptibility. Am J Hum Genet 66(2):494–500

Burdelya LG, Komarova EA, Hill JE, Browder T, Tararova ND, Mavrakis L, DiCorleto PE, Folkman J, Gudkov AV (2006) Inhibition of p53 response in tumor stroma improves efficacy of anticancer treatment by increasing antiangiogenic effects of chemotherapy and radiotherapy in mice. Cancer Res 66(19):9356–9361

Burdelya LG, Gleiberman AS, Toshkov I, Aygun-Sunar S, Bapardekar M, Manderscheid-Kern P, Bellnier D, Krivokrysenko VI, Feinstein E, Gudkov AV (2012) Toll-like receptor 5 agonist protects mice from dermatitis and oral mucositis caused by local radiation: implications for head-and-neck cancer radiotherapy. Int J Radiat Oncol Biol Phys 83(1):228–234

Cerri M, Tinganelli W, Negrini M, Helm A, Scifoni E, Tommasino F, Sioli M, Zoccoli A, Durante M (2016) Hibernation for space travel: impact on radioprotection. Life Sci Space Res (Amst) 11:1–9

Cronkite EP (1964) The diagnosis, treatment, and prognosis of human radiation injury from whole-body exposure. Ann NY Acad Sci 114:341–355

Davis PA, Polagruto JA, Valacchi G, Phung A, Soucek K, Keen CL, Gershwin ME (2006) Effect of apple extracts on NF-kappaB activation in human umbilical vein endothelial cells. Exp Biol Med (Maywood) 231(5):594–598

Donnadieu-Claraz M, Benderitter M, Joubert C, Voisin P (1999) Biochemical indicators of whole-body gamma-radiation effects in the pig. Int J Radiat Biol 75(2):165–174

Drouet M, Herodin F (2010) Radiation victim management and the haematologist in the future: time to revisit therapeutic guidelines? Int J Radiat Biol 86(8):636–648

Durante M, Cucinotta FA (2008) Heavy ion carcinogenesis and human space exploration. Nat Rev Cancer 8(6):465–472

Ferlazzo ML, Bourguignon M, Foray N (2017) Functional assays for individual radiosensitivity: a critical review. Seminar Radiat Oncol 27(4):310–315

Foray N, Bourguignon M, Hamada N (2016) Individual response to ionizing radiation. Mutat Res 770(Pt B):369–386

Genkinger JM, Platz EA, Hoffman SC, Comstock GW, Helzlsouer KJ (2004) Fruit, vegetable, and antioxidant intake and all-cause, cancer, and cardiovascular disease mortality in a community-dwelling population in Washington County, Maryland. Am J Epidemiol 160(12):1223–1233

Guan J, Wan XS, Zhou Z, Ware J, Donahue JJ, Biaglow JE, Kennedy AR (2004) Effects of dietary supplements on space radiation-induced oxidative stress in Sprague-Dawley rats. Radiat Res 162(5):572–579

Gudkov AV, Komarova EA (2010) Radioprotection: smart games with death. J Clin Invest 120(7):2270–2273

Halliwell B (2000) The antioxidant paradox. Lancet 355(9210):1179–1180

Halliwell B (2013) The antioxidant paradox: less paradoxical now? Br J Clin Pharmacol 75(3):637–644

Huang C, Zhang D, Li J, Tong Q, Stoner GD (2007) Differential inhibition of UV-induced activation of NF kappa B and AP-1 by extracts from black raspberries, strawberries, and blueberries. Nutr Cancer 58(2):205–212

Inano H, Onoda M, Inafuku N, Kubota M, Kamada Y, Osawa T, Kobayashi H, Wakabayashi K (2000) Potent preventive action of curcumin on radiation-induced initiation of mammary tumorigenesis in rats. Carcinogenesis 21(10):1835–1841

Copernic Project Investigators, Granzotto A, Benadjaoud MA, Vogin G, Devic C, Ferlazzo ML, Bodgi L, Pereira S, Sonzogni L, Forcheron F, Viau M, Etaix A, Malek K, Mengue-Bindjeme L, Escoffier C, Rouvet I, Zabot M-T, Joubert A, Vincent A, Venezia ND, Bourguignon M, Canat E-P, d'Hombres A, Thébaud E, Orbach D, Stoppa-Lyonnet D, Radji A, Doré E, Pointreau Y, Bourgier C, Leblond P, Defachelles A-S, Lervat C, Guey S, Feuvret L, Gilsoul F, Berger C,

Moncharmont C, de Laroche G, Moreau-Claeys M-V, Chavaudra N, Combemale P, Biston M-C, Malet C, Martel-Lafay I, Laude C, Hau-Desbat N-H, Ziouéche A, Tanguy R, Sunyach M-P, Racadot S, Pommier P, Claude L, Baleydier F, Fleury B, de Crevoisier R, Simon J-M, Verrelle P, Peiffert D, Belkacemi Y, Bourhis J, Lartigau E, Carrie C, De Vathaire F, Eschwege F, Puisieux A, Lagrange J-L, Balosso J, Foray N (2016) Influence of Nucleoshuttling of the ATM protein in the healthy tissues response to radiation therapy: toward a molecular classification of human Radiosensitivity. Int J Radiat Oncol Biol Phys 94(3):450–460

Jacob A, Shah KG, Wu R, Wang P (2010) Ghrelin as a novel therapy for radiation combined injury. Mol Med 16(3–4):137–143

Kamran MZ, Ranjan A, Kaur N, Sur S, Tandon V (2016) Radioprotective agents: strategies and translational advances. Med Res Rev 36(3):461–493

Kennedy AR (2014) Biological effects of space radiation and development of effective counter-measures. Life Sci Space Res (Amst) 1:10–43

Kiang JG, Olabisi AO (2019) Radiation: a poly-traumatic hit leading to multi-organ injury. Cell Biosci 9:25

Kleiman NJ, David J, Elliston CD, Hopkins KM, Smilenov LB, Brenner DJ, Worgul BV, Hall EJ, Lieberman HB (2007) Mrad9 and atm haploinsufficiency enhance spontaneous and X-ray-induced cataractogenesis in mice. Radiat Res 168(5):567–573

Komarov PG, Komarova EA, Kondratov RV, Christov-Tselkov K, Coon JS, Chernov MV, Gudkov AV (1999) A chemical inhibitor of p53 that protects mice from the side effects of cancer therapy. Science 285(5434):1733–1737

Kouvaris JR, Kouloulias VE, Vlahos LJ (2007) Amifostine: the first selective-target and broad-spectrum radioprotector. Oncologist 12(6):738–747

Kresty LA, Morse MA, Morgan C, Carlton PS, Lu J, Gupta A, Blackwood M, Stoner GD (2001) Chemoprevention of esophageal tumorigenesis by dietary administration of lyophilized black raspberries. Cancer Res 61(16):6112–6119

Krukowski K, Grue K, Frias ES, Pietrykowski J, Jones T, Nelson G, Rosi S (2018) Female mice are protected from space radiation-induced maladaptive responses. Brain Behav Immun 74:106–120

MacVittie TJ, Farese AM, Jackson W III (2005) Defining the full therapeutic potential of recombinant growth factors in the post radiation-accident environment: the effect of supportive care plus administration of G-CSF. Health Phys 89(5):546–555

Madhusoodhanan R, Natarajan M, Singh JV, Jamgade A, Awasthi V, Anant S, Herman TS, Aravindan N (2010) Effect of black raspberry extract in inhibiting NFkappa B dependent radioprotection in human breast cancer cells. Nutr Cancer 62(1):93–104

Narendran N, Luzhna L, Kovalchuk O (2019) Sex difference of radiation response in occupational and accidental exposure. Front Genet 10:260

NASA (2015) TA 6: human health, life support, and habitation systems. NASA Technology Roadmaps. National Aeronautics and Space Administration (NASA), Washington, DC

Neal R, Matthews RH, Lutz P, Ercal N (2003) Antioxidant role of N-acetyl cysteine isomers following high dose irradiation. Free Radic Biol Med 34(6):689–695

Poulose SM, Rabin BM, Bielinski DF, Kelly ME, Miller MG, Thanthaeng N, Shukitt-Hale B (2017) Neurochemical differences in learning and memory paradigms among rats supplemented with anthocyanin-rich blueberry diets and exposed to acute doses of (56)Fe particles. Life Sci Space Res (Amst) 12:16–23

Rabin BM, Shukitt-Hale B, Joseph J, Todd P (2005a) Diet as a factor in behavioral radiation protection following exposure to heavy particles. Gravit Space Biol Bull 18(2):71–77

Rabin BM, Joseph JA, Shukitt-Hale B (2005b) Effects of age and diet on the heavy particle-induced disruption of operant responding produced by a ground-based model for exposure to cosmic rays. Brain Res 1036(1–2):122–129

Romero-Weaver AL, Ni J, Lin L, Kennedy AR (2014) Orally administered fructose increases the numbers of peripheral lymphocytes reduced by exposure of mice to gamma or SPE-like proton radiation. Life Sci Space Res (Amst) 2:80–85

Sasse AD, Clark LG, Sasse EC, Clark OA (2006) Amifostine reduces side effects and improves complete response rate during radiotherapy: results of a meta-analysis. Int J Radiat Oncol Biol Phys 64(3):784–791

Sauvaget C, Kasagi F, Waldren CA (2004) Dietary factors and cancer mortality among atomic-bomb survivors. Mutat Res 551(1–2):145–152

Schreurs AS, Shirazi-Fard Y, Shahnazari M, Alwood JS, Truong TA, Tahimic CG, Limoli CL, Turner ND, Halloran B, Globus RK (2016) Dried plum diet protects from bone loss caused by ionizing radiation. Sci Rep 6:21343

Shukitt-Hale B, Carey AN, Jenkins D, Rabin BM, Joseph JA (2007) Beneficial effects of fruit extracts on neuronal function and behavior in a rodent model of accelerated aging. Neurobiol Aging 28(8):1187–1194

Singh VK, Seed TM (2019) The efficacy and safety of amifostine for the acute radiation syndrome. Expert Opin Drug Saf 18(11):1077–1090

Smith SM, Zwart SR, Block G, Rice BL, Davis-Street JE (2005) The nutritional status of astronauts is altered after long-term space flight aboard the international Space Station. J Nutr 135(3):437–443

Sridharan DM, Asaithamby A, Blattnig SR, Costes SV, Doetsch PW, Dynan WS, Hahnfeldt P, Hlatky L, Kidane Y, Kronenberg A, Naidu MD, Peterson LE, Plante I, Ponomarev AL, Saha J, Snijders AM, Srinivasan K, Tang J, Werner E, Pluth JM (2016) Evaluating biomarkers to model cancer risk post cosmic ray exposure. Life Sci Space Res (Amst) 9:19–47

Taylor AM, Harnden DG, Arlett CF, Harcourt SA, Lehmann AR, Stevens S, Bridges BA (1975) Ataxia telangiectasia: a human mutation with abnormal radiation sensitivity. Nature 258(5534):427–429

Thompson D, Duedal S, Kirner J, McGuffog L, Last J, Reiman A, Byrd P, Taylor M, Easton DF (2005) Cancer risks and mortality in heterozygous ATM mutation carriers. J Natl Cancer Inst 97(11):813–822

Tinganelli W, Hitrec T, Romani F, Simoniello P, Squarcio F, Stanzani A, Piscitiello E, Marchesano V, Luppi M, Sioli M, Helm A, Compagnone G, Morganti AG, Amici R, Negrini M, Zoccoli A, Durante M, Cerri M (2019) Hibernation and radioprotection: gene expression in the liver and testicle of rats irradiated under synthetic torpor. Int J Mol Sci 20(2):352

Tomonaga M (1962) Leukaemia in Nagasaki atomic bomb survivors from 1945 through 1959. Bull World Health Organ 26:619–631

Toshkov IA, Gleiberman AS, Mett VL, Hutson AD, Singh AK, Gudkov AV, Burdelya LG (2017) Mitigation of radiation-induced epithelial damage by the TLR5 agonist entolimod in a mouse model of fractionated head and neck irradiation. Radiat Res 187(5):570–580

Weiss JF, Landauer MR (2000) Radioprotection by antioxidants. Ann NY Acad Sci 899:44–60

Wong ET, Tergaonkar V (2009) Roles of NF-kappaB in health and disease: mechanisms and therapeutic potential. Clin Sci (Lond) 116(6):451–465

Worgul BV, Smilenov L, Brenner DJ, Junk A, Zhou W, Hall EJ (2002) Atm heterozygous mice are more sensitive to radiation-induced cataracts than are their wild-type counterparts. Proc Natl Acad Sci USA 99(15):9836–9839

Yamamoto Y, Gaynor RB (2001) Therapeutic potential of inhibition of the NF-kappaB pathway in the treatment of inflammation and cancer. J Clin Invest 107(2):135–142

Chapter 6
Challenges for Exploratory Missions

Abstract Passive and active dosimeters for space missions are available and constantly improved. The radiation environment of the International Space Station (ISS) is well characterized and radiation detectors are measuring on the surface of Mars and Moon. Concepts of storm shelters are implemented in the upcoming Artemis missions to the Moon. The first human phantom experiment beyond LEO will bring about the depth dose distribution in the female body. Space radiobiological research currently shifts from single ion experiments to combinations of several ions and from classical radiobiological endpoints such as cellular survival and chromosomal aberrations to tissue- and organ-specific responses that are relevant for degenerative diseases. Furthermore, in addition to monolayer cell cultures, more complex model systems such as organoids are used for accelerator-based research. Other spaceflight environmental factors than microgravity—stress, isolation, sleep deprivation—are coming into the research focus for combined effects. For countermeasure development, tests with well-known drugs are suggested.

Keywords Dosimetry for exploration missions · Shielding · Spaceflight environmental factors · Space radiation biology research

6.1 Dosimetry and Shielding: Are We Ready for Launch?

Long-term exploration missions will come, be it either to the Moon, to near-Earth asteroids or in the really long term to Mars. We now understand our radiation environment in free space much better than in the beginning of the space age, where

© The Author(s), under exclusive licence to Springer Nature Switzerland AG 2020
C. E. Hellweg et al., *Radiation in Space: Relevance and Risk for Human Missions*, SpringerBriefs in Space Life Sciences,
https://doi.org/10.1007/978-3-030-46744-9_6

also the problem of radiation dosimetry and shielding was just at the start of discussion.

We know that with relevant shielding—as a storm shelter—applied we can reduce the dose due to a Solar Particle Event (SPE) to a relevant exposure which will be beneath the relevant limits and thereby also decrease the risk of radiation thickness. We know that by choosing the time of flight of a relevant real long-term mission (as for example at solar maximum) we can decrease the dose due to GCR and thereby also decrease the Risk of Exposure-Induced Death (REID). In the last 10 years, we landed with radiation detectors on Mars (MSL-RAD) and on the far side of the Moon (LND). We increased the number of radiation instruments onboard the ISS, and we are developing new detector systems for the upcoming Orion missions, which will bring back astronauts at the long term on the surface of the Moon. Furthermore, on the NASA Artemis I mission, the MATROSHKA AstroRad Radiation Experiment (MARE) will for the first time fly two female phantoms on its way to the Moon to determine the depth dose distribution as essential prerequisite for risk estimation. Both phantoms will be equipped with radiation detectors and one will wear a radiation protection vest. Both are modeled on humans, so that the radiation dose can be measured in the particularly radiation-sensitive organs. We also use these data from the different instruments and experiments to benchmark and to further develop our radiation transport models and codes, being able to better predict the radiation exposure of upcoming missions. Having all this in mind—there is still a lot to do, a lot to measure, a lot do discover, but we are on a good way.

6.2   Open Questions in Space Radiation Biology and Risk Assessment: Can We Tame the Radiation Risk?

The picture of the molecular and cellular changes after a heavy ion hit becomes continuously clearer. Depending on the radiation dose and the dose rate, certain cellular outcomes will probably be favored in sustaining the astronauts' health. During an acute high-dose and high-dose rate exposure by an SPE, survival of as many cells as possible is desirable in order to avoid organ failure. Concerning the long-term risk of low-dose and low-dose rate exposure to GCR, the death of any cell with remaining DNA damage, mutations or chromosomal aberrations, could help to reduce the cancer risk attributed to space radiation. On the other hand, in organs with low cell turnover, low regeneration potential and a high percentage of terminally differentiated cells such as the brain, the loss of neurons by cell death might be detrimental and result in continuous accumulation of tissue damage. More research is needed to develop strategies of pharmacologically directing the cellular outcome of space radiation-exposed cells.

The way to a better risk assessment of space radiation effects reducing the uncertainty of cancer induction risk is still long. It is difficult because not only low doses and low dose rates have to be considered, but the radiation quality in space is unique and extremely complex, and the involvement of opposing pathways towards survival or death which can be either beneficial or detrimental ("only a dead cell is a good cell" to prevent cancerogenesis). Therefore, the outcome for an individually hit cell cannot be predicted. Furthermore, most radiobiological experiments were performed with one single ion, with 1 GeV/n iron ions being the most popular. To take the complexity of the space radiation field into account, experiments with two or more ions (protons, helium, and a heavier ion) are performed. Since shortly, a galactic cosmic ray simulation is available at the NASA Space Radiation Laboratory (NSRL) at Brookhaven, Upton, NY, USA (Norbury et al. 2016). As the setup of the GCR simulation takes several weeks, beamtime is very limited and only granted to experiments for which the biological responses to every single ion are known to allow comparison of the single ion to the combined beam results. A different GCR simulation is planned at the new accelerator FAIR at the GSI Helmholtzzentrum für Schwerionenforschung GmbH in Darmstadt, Germany.

There are many knowledge gaps in the connection between well-described early effects and also well-documented late effects. Experiments with three-dimensional cell models, organoids, and organ cultures might help to close this gap. Also, further epidemiological analyses of populations with both low dose exposures and with low- and high-LET exposures should continue to improve risk prediction for space radiation-induced cancer.

Furthermore, space radiation biology research attempts to understand the combined effects of different spaceflight environmental factors, e.g., radiation exposure in combination with an elevated stress level or (simulated) microgravity or even sleep deprivation or isolation. Whether the DNA damage response, especially DNA repair, is altered under microgravity, was investigated with conflicting results, and changes in proliferation under microgravity are suggested as possible explanation for an altered DNA damage response (Moreno-Villanueva et al. 2017) (Sect. 3.2). New research opportunities might become available with an X-ray source on the International Space Station (ISS) and the fluorescence microscope FLUMIAS (https://www.dlr.de/rd/en/desktopdefault.aspx/tabid-2283/3420_read-52416/).

To simulate stress in cell culture, application of corticoids is tested. Such experiments are performed within the current ESA program "Investigations into Biological Effects of Radiation (IBER) Using the GSI Accelerator Facility" (https://www.gsi.de/work/forsc-hung/biophysik/esa_iber.htm) which provides beamtime using high energy (up to 1 GeV/n) ions for research on the biological effects of space radiation. FAIR will expand the research possibilities up to energies of 10 GeV/n. Also the medium energy ions (up to 95 MeV/n) available at the Grand Accélérateur National d'Ions Lourds (GANIL) provide LET at cellular level that are highly relevant for space radiation research.

Currently, from all space agencies, only NASA has developed a sophisticated space radiation risk model. Other agencies have set exposure limits without considering the population background, sex and age of the astronauts and without model calculations. An international agreement on space radiation risk assessment and exposure limits is needed for international exploration missions.

Concerning the Acute Radiation Syndrome (ARS), NASA mostly stopped the research as this condition can be prevented by a warning system in combination with a radiation shelter. Also, ARS research using animal models is stressful and the chances to find a better cure are low. Ethical considerations on ARS research for spaceflight using animals might therefore result in the conclusion that the animal suffering is not justified.

6.3   Countermeasures: Do We Know Enough?

Finding a magic pill that protects from cancer and even aging is humankind's dream. There is abundant scientific literature on "magic" phytochemicals, including resveratrol, curcumin, and green tea polyphenols. In the end, none of these substances keeps its promises and the spontaneous cancer rate remains high. Some rodent experiments showed striking effects for berries, but it is unclear to what extent this effect can be translated to humans for a space mission. To speed up the process of countermeasure development, NASA suggests testing well-known pharmaceuticals such as aspirin because their risk and action profiles are well known.

Based on the assumption that cancer is a multistep process that requires several genetic events that can be induced by different carcinogens, a recommendation to avoid other carcinogens such as tobacco smoke can be given. Generally, a healthy lifestyle with a diet rich in vegetables, fruits, and fiber and at least 150 min of moderate-intensity aerobic activity per week is recommended as cancer prophylaxis.

6.4   Benefits of Space Radiation Research for Terrestrial Applications

Beyond the direct benefit of space radiation research for radiation protection during human space missions, this research contributes to the understanding of side effects of tumor radiotherapy using accelerated ions, e.g., protons and carbon ions. Tumor therapy aims at eliminating all cells of a tumor while sparing the surrounding normal tissue. The depth dose distribution and the minor lateral expansion of energetic ion beams are highly favorable to achieve this goal (Kraft 1990): In the entrance channel, the particles have a high energy, and the absorbed dose is low. Traversing the tissue, the particle increasingly loses energy and the energy deposition reaches its maximum shortly before the particle comes to a complete stop (Bragg peak of the range versus LET curve, Fig. 3.3). Beyond this maximum, only a very small physical dose is delivered to the tissue (Kraft et al. 1999) from secondary charged particles and neutrons (but with high RBE).

The use of high-LET carbon ions instead of low-LET radiation, such as X-rays and γ-rays, bears further advantages (Kraft 1990). While protons have a relative biological effectiveness (RBE, Sect. 3.3) comparable to or only slightly higher than that of photons (Courdi et al. 1994; Gueulette et al. 1997), carbon ions used for therapy have a higher RBE of ~3 (Kanai et al. 1997; Ando et al. 2006).

The energy of the carbon ion beam is modulated in such a way that the Bragg peak (Fig. 3.3) is located within the tumor volume. The Bragg peak is spread out by weighted superposition of elementary Bragg curves to cover the whole tumor with a uniform biological dose (Bortfeld and Schlegel 1996; Jette and Chen 2011). In clinical use of proton and carbon ion beams, a deep-seated tumor is irradiated with this Spread-Out Bragg Peak (SOBP) (Coutrakon et al. 1991) displaying a high LET, whereas surface skin is exposed to the lower doses of the entrance plateau (Courdi et al. 1994).

For ion beam therapy, ~ 300 MeV protons and up to 450 MeV/n carbon ions are used (Jäkel 2008). In the entrance region of a carbon ion beam, the LET is around 10 keV/µm, and it rises to 50–80 keV/µm in the Bragg peak region (Durante and Loeffler 2010). These energies and LET ranges are also highly relevant in space radiation research. Therefore, the understanding of basic mechanisms of complex DNA damage induction and repair, DNA damage response, cell killing, cell survival, and other cell fates after heavy ion exposure is as relevant for space radiobiology as for particle therapy of tumors (Hellweg et al. 2018; Ray et al. 2018). Additionally, the role of the genetic background (e.g., p53 gene status (Mori et al. 2009) for the cellular outcomes is important to understand on the one hand, possible tumor radioresistance, and on the other hand, inter-individual variability of sensivitity towards heavy ion exposure, enabling personalized approaches for space radiation protection and tumor therapy.

One drawback of carbon ion therapy might be the fact that despite the surrounding tissues are hit by much lower doses than in conventional radiotherapy, the RBE at such doses might be high for inducing late effects. The data from clinical studies are currently too heterogeneous to draw a definite conclusion for the risk of late effects for particle therapy (Suit et al. 2007). Here, the numerous attempts to determine the tumor induction RBE of different charged particles in space radiobiology (Sect. 3.3) could support assessing the risks of particle therapy to trigger late sequelae.

Both, space radiation research and particle therapy could profit from successful countermeasure development. A systemic countermeasure would be preferable for space missions, while a local approach could further protect healthy, (albeit small) tumor-surrounding tissue that is hit by charged particles from acute or late effects. Identification of radioprotective pathways bears the possibility to trigger or enhance these pathways as well as to suppress them. Therefore, based on such results, inhibitors of radioprotective pathways could be developed in order to further augment the tumor cell killing effect of carbon ions.

After closing heavy ion radiotherapy at the Lawrence Berkeley National Laboratory (LBNL) in 1992 (Castro et al. 1994), the Heavy Ion Medical Accelerator in Chiba (HIMAC) at the National Institute of Radiological Sciences (NIRS) in Japan (Hirao et al. 2002), and the GSI ("Gesellschaft für Schwerionenforschung" now "GSI Helmholtzzentrum für Schwerionenforschung") in Darmstadt, Germany, formed the nucleus for carbon ion radiotherapy (Kraft et al. 1994). Based on the pioneering work at GSI, patients are now treated with carbon ion radiotherapy at the Heidelberg Ion Therapy (HIT) facility in Heidelberg. In 2017, 72 particle therapy facilities were in operation worldwide, and a number of 140 was expected to be reached in 2021 (Dosanjh et al. 2018a). Partly, these facilities are open to radiobiological research (Dosanjh et al. 2018b), thereby offering opportunities to close the gaps in the understanding of biological effects of charged particles.

6.5 Conclusion

Astronauts on exploration missions should be equipped with appropriate personal passive and active dosimeters. They should have access to a radiation shelter and to information on space weather especially upcoming SPE. The spaceship should be optimized for GCR shielding, especially the sleeping quarter, and should be equipped with efficient propulsion to reduce the time spent in deep space. Habitats on celestial bodies should provide better GCR shielding to reduce the mission dose as much as possible. The diet should contain fruits, berries, vegetables, and fiber. Astronauts should be given the opportunity to participate in an up-to-date space radiation protection course. The space pharmacy should be equipped with drugs against ARS for scenarios of a worst case SPE combined with technical failure of the warning system. From the point of view of healthy lifetime reduction, the recruitment of older crewmembers is recommended. It remains open whether all these measures will reduce the disease risks to an acceptable extent using the currently available model. In the end, it remains a decision of the involved societies and space agencies which space radiation risk is acceptable for space exploration missions and an informed consent of astronauts is required. The ongoing space radiation research is contributing to humankind's continuous search for remedy against aging and cancer.

References

Ando K, Koike S, Uzawa A, Takai N, Fukawa T, Furusawa Y, Aoki M, Hirayama R (2006) Repair of skin damage during fractionated irradiation with gamma rays and low-LET carbon ions. J Radiat Res 47(2):167–174

Bortfeld T, Schlegel W (1996) An analytical approximation of depth-dose distributions for therapeutic proton beams. Phys Med Biol 41(8):1331–1339

Castro JR, Linstadt DE, Bahary JP, Petti PL, Daftari I, Collier JM, Gutin PH, Gauger G, Phillips TL (1994) Experience in charged particle irradiation of tumors of the skull base: 1977–1992. Int J Radiat Oncol Biol Phys 29(4):647–655

Courdi A, Brassart N, Herault J, Chauvel P (1994) The depth-dependent radiation response of human melanoma cells exposed to 65 MeV protons. Br J Radiol 67(800):800–804

Coutrakon G, Bauman M, Lesyna D, Miller D, Nusbaum J, Slater J, Johanning J, Miranda J, DeLuca PM Jr, Siebers J et al (1991) A prototype beam delivery system for the proton medical accelerator at Loma Linda. Med Phys 18(6):1093–1099

Dosanjh M, Amaldi U, Mayer R, Poetter R (2018a) ENLIGHT: European network for light ion hadron therapy. Radiother Oncol 128(1):76–82

Dosanjh M, Jones B, Pawelke J, Pruschy M, Sorensen BS (2018b) Overview of research and therapy facilities for radiobiological experimental work in particle therapy. Report from the European particle therapy network radiobiology group. Radiother Oncol 128(1):14–18

Durante M, Loeffler JS (2010) Charged particles in radiation oncology. Nat Rev Clin Oncol 7(1):37–43

Gueulette J, Bohm L, De Coster BM, Vynckier S, Octave-Prignot M, Schreuder AN, Symons JE, Jones DT, Wambersie A, Scalliet P (1997) RBE variation as a function of depth in the 200-MeV proton beam produced at the National Accelerator Centre in Faure (South Africa). Radiother Oncol 42(3):303–309

Hellweg CE, Chishti AA, Diegeler D, Spitta LF, Henschenmacher B, Baumstark-Khan C (2018) Molecular Signaling in Response to Charged Particle Exposures and its Importance in Particle Therapy. Int J Part Ther 5(1):60–73

Hirao Y, Ogawa H, Yamada S, Sato Y, Yamada T, Sato K, Itano A, Kanazawa M, Noda K, Kawachi K, Endo M, Kanai T, Kohno T, Sudou M, Minohara S, Kitagawa A, Soga F, Takada E, Watanabe S, Endo K, Kumada M, Matsumoto S (2002) Heavy ion synchrotron for medical use —HIMAC project at NIRS-Japan. Nucl Phys A 538:541–550

Jäkel O (2008) The relative biological effectiveness of proton and ion beams. Z Med Phys 18(4):276–285

Jette D, Chen W (2011) Creating a spread-out Bragg peak in proton beams. Phys Med Biol 56(11):N131–N138

Kanai T, Furusawa Y, Fukutsu K, Itsukaichi H, Eguchi-Kasai K, Ohara H (1997) Irradiation of mixed beam and design of spread-out Bragg peak for heavy-ion radiotherapy. Radiat Res 147(1):78–85

Kraft G (1990) The radiobiological and physical basis for radiotherapy with protons and heavier ions. Strahlenther Onkol 166(1):10–13

Kraft G, Becher W, Blasche K, Böhne D, Franczak B, Haberer T, Kraft-Weyrather W, Kraemer M, Langenbeck B, Lenz G, Ritter S, Scholz M, Schardt D, Stelzer H, Strehl P, Weber U (1994) The Darmstadt program HITAG: heavy ion therapy at GSI. In: Amaldi U, Larsson B (eds) Hadrontherapy in oncology. Elsevier Science Pub Co., Amsterdam, NL, pp 217–228

Kraft G, Scholz M, Bechthold U (1999) Tumor therapy and track structure. Radiat Environ Biophys 38(4):229–237

Moreno-Villanueva M, Wong M, Lu T, Zhang Y, Wu H (2017) Interplay of space radiation and microgravity in DNA damage and DNA damage response. NPJ Microgravity 3:14

Mori E, Takahashi A, Yamakawa N, Kirita T, Ohnishi T (2009) High LET heavy ion radiation induces p53-independent apoptosis. J Radiat Res 50(1):37–42

Norbury JW, Schimmerling W, Slaba TC, Azzam EI, Badavi FF, Baiocco G, Benton E, Bindi V, Blakely EA, Blattnig SR, Boothman DA, Borak TB, Britten RA, Curtis S, Dingfelder M, Durante M, Dynan WS, Eisch AJ, Robin Elgart S, Goodhead DT, Guida PM, Heilbronn LH, Hellweg CE, Huff JL, Kronenberg A, La Tessa C, Lowenstein DI, Miller J, Morita T, Narici L, Nelson GA, Norman RB, Ottolenghi A, Patel ZS, Reitz G, Rusek A, Schreurs AS, Scott-Carnell LA, Semones E, Shay JW, Shurshakov VA, Sihver L, Simonsen LC, Story MD, Turker MS, Uchihori Y, Williams J, Zeitlin CJ (2016) Galactic cosmic ray simulation at the NASA space radiation laboratory. Life Sci Space Res (Amst) 8:38–51

Ray S, Cekanaviciute E, Lima IP, Sorensen BS, Costes SV (2018) Comparing photon and charged particle therapy using dna damage biomarkers. Int J Part Ther 5(1):15–24

Suit H, Goldberg S, Niemierko A, Ancukiewicz M, Hall E, Goitein M, Wong W, Paganetti H (2007) Secondary carcinogenesis in patients treated with radiation: a review of data on radiation-induced cancers in human, non-human primate, canine and rodent subjects. Radiat Res 167(1):12–42

Printed in the United States
By Bookmasters